Multiscale Fatigue Modelling of Metals

M. Mlikota, S. Schmauder, Ž. Božić

Edited by

K.J. Dogahe

Institute for Materials Testing, Materials Science and Strength of Materials (IMWF), University of Stuttgart, 70569 Stuttgart, Germany

kiarash.dogahe@imwf.uni-stuttgart.de

Copyright © 2022 by the authors

Published by **Materials Research Forum LLC**
Millersville, PA 17551, USA

Published as part of the book series
Materials Research Foundations
Volume 114 (2022)
ISSN 2471-8890 (Print)
ISSN 2471-8904 (Online)

Print ISBN 978-1-64490-164-9
eBook ISBN 978-1-64490-165-6

Distributed worldwide by

Materials Research Forum LLC
105 Springdale Lane
Millersville, PA 17551
USA
https://www.mrforum.com

Manufactured in the United States of America
10 9 8 7 6 5 4 3 2 1

Table of Contents

Preface

Fatigue is one of the most important and conventional failure modes for engineering components. The significance of this phenomenon is that the failure may occur even when the stresses in the critical regions are below the elastic limit. In order to thoroughly estimate the fatigue life of a metallic component, it is necessary to start the investigation from the very beginning steps of fatigue. The very fundamental stage during the course of fatigue is the initiation of fatigue microcracks on slip bands within the microstructure. For some alloys the crack initiation step can acquire even up to 90% of the total fatigue life of the component. Nevertheless, it is worth mentioning that engineering materials perform various behavior during the different fatigue stages. Due to this fact a multiscale approach is required for modelling the whole fatigue life of a component.

The main aim of this book is the multiscale modelling of metallic materials performance during the different stages of the fatigue life, as well as the investigation of the involved parameters for each stage.

The first chapter of this book is about "The numerical determination of Paris law constants for carbon steel using a two-scale model". The Paris law represents the correlation between the range of stress intensity factor (ΔK) and fatigue crack growth rate (da/dN) under the impression of the material constants C and m. In order to avoid expensive experimental tests for the determination of Paris law constants C and m for carbon steel, a two-scale method is employed in this chapter. In this regard, to calculate the fatigue crack growth rate at the crack tip, by using the Tanaka-Mura (TM) equation, a microstructural model is generated at this area. A macro-model is employed as well to calculate the stress intensity factor. With this micro-macro approach, a correlation between the stress intensity factor and the crack growth rate for the specified crack lengths comes in hand which is used to determine the Paris law constants.

Chapter 2 is about the "Calculation of the Wöhler (S-N) curve using a two-scale model". In this chapter the estimation of the complete fatigue life of carbon steel under cyclic loading is represented by the Wöhler (S-N) curve. This estimation is achieved by modelling the initiation of a short crack and subsequent growth of the long crack. In this regard, a two scale model which includes the microstructure of the carbon steel is developed using the Finite Element Method, in order to obtain the grain based stress distribution. The required number of cycles for crack initiation based on this nonuniform stress distribution is calculated by employing again the Tanaka-Mura model. Finally, the analysis of the long crack growth is performed using the Paris law.

The 3rd chapter "On the critical resolved shear stress and its importance in the fatigue performance of steels and other metals with different crystallographic structures", is dealing with the numerical estimation of the fatigue life in the form of Wöhler *S-N* curves of metals with different crystallographic structures. Again, in this chapter, the whole fatigue life of the material is divided in two steps, which includes fatigue micro-crack initiation which is modelled by a TM formulation followed by long crack growth analysis which is conducted by using the Paris law. It is shown that the fatigue life curve of metals is not predominately determined by their crystallographic structures, but it is determined by the material parameter which is known as critical resolved shear stress (CRSS).

The book closes with chapter 4 with the title: "A newly discovered relation between the critical resolved shear stress and fatigue endurance limit for metallic materials". This chapter introduces a better understanding of the fatigue process of metallic materials by specifying the correlation between the fatigue strength and the critical resolved shear stress (CRSS) which is an intrinsic property of metallic materials. In this respect a multiscale approach of fatigue modelling is employed in order to correlate the endurance limit to the CRSS rather than the ultimate strength, as often done in the past.

Contributions of the working group at the Institute for Materials Testing, Materials Science and Strength of Materials (IMWF) in Stuttgart and the Faculty of Mechanical Engineering and Naval Architecture at the University of Zagreb can be found in the first two chapters.

With the contained articles in this book, the significance of multiscale modelling in terms of fatigue performance of metallic materials is shown. We hope that the content of this book being inspiring for the readers and helps to develop new ideas.

Stuttgart, October 2021

Siegfried Schmauder

Kiarash J. Dogahe

Marijo Mlikota

Željko Božić

Multiscale Fatigue Modelling of Metals   Materials Research Forum LLC
Materials Research Foundations **114** (2022) 1-15   https://doi.org/10.21741/9781644901656-1

Chapter 1

Numerical Determination of Paris Law Constants for Carbon Steel Using a Two-Scale Model

M. Mlikota[1], S. Staib[2], S. Schmauder[1], Ž. Božić[2]

[1] Institute for Materials Testing, Materials Science and Strength of Materials (IMWF), University of Stuttgart, Pfaffenwaldring 32, 70569 Stuttgart, Germany

[2] Faculty of Mechanical Engineering and Naval Architecture, University of Zagreb, I. Lučića 5, 10000 Zagreb, Croatia

Abstract

For most engineering alloys, the long fatigue crack growth under a certain stress level can be described by the Paris law. The law provides a correlation between the fatigue crack growth rate (FCGR or da/dN), the range of stress intensity factor (ΔK), and the material constants C and m. A well-established test procedure is typically used to determine the Paris law constants C and m, considering standard specimens, notched and pre-cracked. Definition of all the details necessary to obtain feasible and comparable Paris law constants are covered by standards. However, these cost-expensive tests can be replaced by appropriate numerical calculations. In this respect, this paper deals with the numerical determination of Paris law constants for carbon steel using a two-scale model. A micro-model containing the microstructure of a material is generated using the Finite Element Method (FEM) to calculate the fatigue crack growth rate at a crack tip. The model is based on the Tanaka-Mura equation. On the other side, a macro-model serves for the calculation of the stress intensity factor. The analysis yields a relationship between the crack growth rates and the stress intensity factors for defined crack lengths which is then used to determine the Paris law constants.

Keywords

Paris Law, Fatigue Crack Growth Rate (FCGR), Stress Intensity Factor (ΔK), Finite Element Method (FEM), Tanaka-Mura Equation

Originally published in the Journal of Physics: Conference Series, 843 (2017) 012042
https://doi.org/10.1088/1742-6596/843/1/012042

Contents

1.   Introduction

Structural health monitoring and detection of failures are highly important for fitness and service assessment of structures. Failure of structures under fatigue loading can occur at load levels below the yield stress of the used material. Therefore, it is of special importance to be able to make predictions of life cycles until catastrophic fracture occurs. A well-known example of such catastrophic failure is the huge train accident in Eschede where one wheel of the train broke due to cyclic loading (Fig. 1) has not been considered during construction [1]. On the other hand, not every single crack has to be critical immediately. Often structures can withstand cracks up to a certain crack length until unstable fracture occurs. In order to determine the point where crack growth becomes unstable, it is necessary to develop methods to describe crack growth mathematically.

Figure 1. Fracture surface of the broken train wheel [1].

A very common and often used method for the characterization of the long crack growth is the Paris law which gives a relationship between the fatigue crack growth rate (FCGR or da/dN) and the stress intensity factor (ΔK) at the crack tip during the stable crack growth [2], Eq. 1. A typical fatigue growth rate curve – also known as da/dN versus ΔK curve – is shown in Fig. 2. The curve is defined by regions I, II and III. The Paris law relationship can be visualized as a straight line for the region of stable crack growth – region II [3]. The accompanying mathematical equation contains two material parameters C and m, where m represents the slope of the line in Fig. 2 and C the y-axis-intercept [4]:

$$\frac{da}{dN} = C(\Delta K)^m \qquad (1)$$

The crack growth rates in the region II are typically in the order of 10^{-9} to 10^{-6} m/cycle and correspond to stable crack growth. The constants C and m are usually determined in experiments [5-9] and depend on the material and various influencing factors such as temperature, environmental medium and loading ratio [5, 10]. The last is probably the most significant and usually results in closely spaced lines parallel to each other. For metals the exponent is typically of the order m = 2…4 [10].

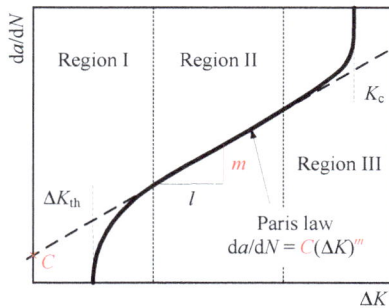

Figure 2. Typical growth behavior of long fatigue cracks [3].

Since the experimental determination of the Paris law constants is typically tedious and time consuming, the objective of this paper is to determine them numerically considering the influence of microstructure on the crack growth rate.

2.   Modelling approach and details

The simulations for determination of the Paris law constants are done with the Finite Element (FE) software ABAQUS and by applying a two-scale model. The model contains the microstructural submodel of a material that is used to calculate the fatigue crack growth rate da/dN at a crack tip and the macro-model that serves for the calculation of the stress intensity factor ΔK. In order to get the typical da/dN versus ΔK curve, the crack growth rate and stress intensity factors for six different crack lengths are determined. The constants are then determined from the interpolated straight line, representing region II in Fig. 2, as the slope m and the y-axis-intercept C.

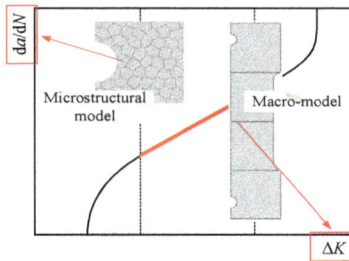

Figure 3. Schematic illustration of the two different approaches.

The input data for the common diagram is calculated using the two aforementioned approaches, as shown in Fig. 3. On the one hand the stress intensity factor is determined using the full-scale macro-model with a center crack and on the basis of Linear Elastic Facture Mechanics (LEFM). The crack growth rate on the other hand is determined using the submodel based on the Tanaka-Mura equation [11, 12]; the submodel is placed in the area behind the crack tip. The grain structure is assigned to the submodel to consider micro-crack initiation processes in the vicinity of the crack tip.

The Tanaka-Mura equation is typically used to estimate the number of cycles that are attributed to the crack nucleation in a single grain [11, 12]. Jezernik *et al.* [13, 14] introduced a modification of the original equation in the sense that the crack does not form instantaneously through the whole grain but it forms in segmental manner:

$$N_s = \frac{8GW_c}{\pi(1-v)d_s(\Delta\bar{\tau} - 2CRSS)^2} \qquad (2)$$

where , N_s is the number of cycles required to form a crack segment in a single grain, i.e. crystal. Furthermore, Eq. 2, contains two microstructure-related parameters, namely the length of slip line segment d_s and the average shear stress range $\Delta\bar{\tau}$ on it. Other material constants (such as shear modulus G, crack initiation energy W_c and Poisson's ratio ν) can be either found in literature for most materials or can alternatively be obtained experimentally. The critical resolved shear stress (*CRSS*) is particularly interesting since it can be obtained by means of Molecular Dynamics (MD) simulations [15]. The *CRSS* represents a critical value of the shear stress along the slip direction that must be overcome for the dislocation to move, i.e. if the magnitude of the resolved shear stress is below the value of the *CRSS*, no dislocation movement is allowed and consequently no pile up at the grain boundary takes place [16].

2.1   Geometry of the macro-models

In order to accomplish the goal of the present paper it was necessary to create two slightly different full-scale macro-models. Both models represent the same geometry, however, the ways of modelling the crack for determination of stress intensity factor ΔK on the one side and fatigue crack growth rate da/dN on the other require certain adjustments at the regions of interest, i.e. on the crack-affected path. More specifically, the macro-model for the calculation of ΔK requires the usage of some special ABAQUS techniques to represent the crack – in this case the seam crack is used – while the macro-model (further: global model) for determination of da/dN needs to be geometrically adjusted to the submodel that is placed at the tip of the structural crack. A seam defines an edge or a face in a model that is originally closed but can open as a crack during an analysis [17].

The models were built on the basis of a specimen with central pre-crack prepared for fatigue testing, shown in Fig. 4, considered in the paper of Božić *et al.* [3]. At the initial state the specimen has a notch and a pre-crack of 2a = 8 mm (Fig. 4, A – Crack detail). Only half of the specimen has to be modeled due to symmetry with respect to the vertical y-axis. The symmetry with respect to the x-axis is aligned with the crack and, therefore, cannot be used in either case; for ΔK determination the applied seam cannot extend along the boundaries of a part and must be embedded within a face of a two-dimensional (2D) part or within a cell of a solid part [17]. In the other case, the used submodel (microstructural model) at the crack tip area requires the transfer of boundary conditions – in this case displacements – from both parts of the global model, the upper and the lower one.

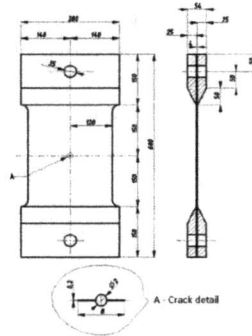

Figure 4. Geometry of the used specimen with a central pre-crack [3].

The structural crack lengths which were used for this study were taken from [3], where ΔK values were determined using the FE software ANSYS. Nevertheless, ΔK values were calculated again – this time with ABAQUS – first to compare the results with previous simulations and second to ensure that the models are built properly. The considered crack lengths for the simulations are listed in Table 1.

Table 1. Considered structural crack lengths.

Crack designation	Crack length (mm)
a_1	9.9
a_2	20.1
a_3	29.9
a_4	39.1
a_5	49.3
a_6	59.3

All models which are used here are built as 2D, in accordance to the geometry of the specimen and the fact that the main central part has a thickness of just 4 mm along the z-direction. The areas in the upper and lower part of the models with relatively higher thickness (54 mm) were also modelled as 2D. Those two parts are used to apply loading conditions and movement constraints, respectively, as shown in Fig.5. The stress state in the middle part of the cracked plate is assumed to be plane stress. The thicknesses of the specimen in both regions, the central and outer, were considered by assigning plane stress thickness to their belonging sections.

As already mentioned only one half of the specimen has to be modelled. The symmetry is realized by using boundary conditions on the vertical y-axis which fix displacements in x-direction as well as rotations of any kind (Fig.5).

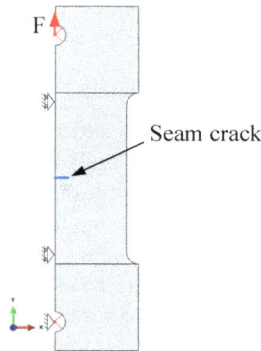

Figure 5. The half of full-scale model with applied boundary and loading conditions, and with the seam crack in the ΔK case.

The material behaviour is assumed to be purely linear elastic; only a small plastic zone at the crack tip is expected (Fig.7), therefore, no plastic material data are necessary to be used. The isotropic material data for the specimen made of conventional mild carbon steel were adopted from the study of Božić *et al.* [3]. The material parameters applied to both models are: Young modulus $E = 206$ GPa, Poisson's ratio $v = 0.3$, yield stress $R_e = 235$ MPa and shear modulus $G = 80$ GPa. It is opportune to introduce the experimentally obtained Paris law constants for the selected material in order to provide validation data for the numerical results that follow later; $m = 2.75$ and $C = 1.43 \times 10^{-11}$ [3].

As the material data, loading and boundary conditions were taken from [3], the testing specimen was exposed to constant amplitude cyclic tension load in a hydraulic fatigue testing machine. The load was applied to the pin which was placed in the hole in the upper part of the model while the pin in the lower hole was fixed (Fig.5). The force range and the stress ratio applied to the half-model are denoted by $F = F_{max} - F_{min}$ and $R = F_{min} / F_{max}$, and were F = 76 800 N and R = 0.0253 [3]. In contrast to the experiments, simulations on the macro-models were performed with static loading conditions, however, in combination with LEFM for the ΔKs determination and with the Tanaka-Mura equation for the determination of da/dN.

For both models continuum plane stress 4-node bilinear elements with reduced integration and hourglass control (CPS4R) were used. In order to deal with the stress singularity at the crack tip in the case of ΔKs calculation, special elements have to be used that are able to indicate the infinite stresses properly. In ABAQUS this is done by collapsing one side of an 8-node isoparametric element connected to the crack tip [17], as can be seen in Fig.7.

2.2 Submodel (microstructural model) details

The submodel or microstructural model is placed right at the tip of the global model structural cracks (Table 2) where their extension is expected. As already mentioned, the Tanaka-Mura equation is typically used to estimate the duration of the short crack initiation stage. In this paper, however, the equation is applied within the microstructural model (Fig.6) with the aim to estimate the rate of the infinitesimal crack extension. The geometry of the submodel can be seen in Fig.6, as well as the location where it is placed with respect to the global model. Its size is selected to 0.4 mm x 0.4 mm including the tip of the structural crack of 0.1 mm radius, which can be identified as a notch. As mentioned before, displacements of the global model are applied to the coinciding edges of the submodel, marked bold in Fig.6.

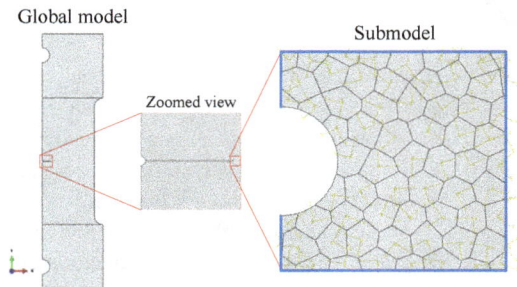

Figure 6. Geometry of the submodel containing a microstructure of the material and its location with respect to the global model.

Moreover, the submodel takes into consideration the microstructure of the investigated material and its influence on the crack growth rate in the initial extension phase. Accordingly, the submodel is partitioned into individual grains using the random Voronoi tesselation. Each individual grain possesses a local coordinate system (Fig.6) and individual slip band orientation. The resulting average grain size, i.e. average slip band length is 50 µm, what is appropriate for the used material.

Concerning the material description, the microstructural model requires orthotropic material data, i.e. elastic constants of cubic crystals. These constants reflect cubic symmetry where $C_{11} = C_{22} = C_{33}$, $C_{12} = C_{13} = C_{23}$ and $C_{44} = C_{55} = C_{66}$, and are calculated using the following equations:

$$C_{11} = \frac{E(1 - v)}{(1 - v - v^2)} \tag{3}$$

$$C_{12} = \frac{E(1 - v)}{(1 - v - 2v^2)} \tag{4}$$

$$C_{44} = G \tag{5}$$

With the isotropic material data from above this gives $C_{11} = 277\ 307$ MPa, $C_{12} = 118\ 846$ MPa and $C_{44} = 80\ 000$ MPa. Furthermore, the microstructural model enriched with the Tanaka-Mura equation requires two additional material properties, namely the critical resolved shear stress ($CRSS$) and the crack initiation energy (W_c). With respect to this, the following values were applied: $CRSS = 117$ MPa [15] and $W_c = 69$ N/mm [18].

The submodel was very fine meshed with the same elements as the global model (CPS4R); what in the end gives smooth stress distribution in the undamaged as well as in the damaged submodel (Fig.9-12). To depict the mesh fineness, a single grain has 120 elements in average.

3   Results

The stress intensity factor ΔK as well as the J-Integral and other fracture mechanics characteristics, can be requested in ABAQUS as history output data. In order to get ΔK for each individual crack of six in total from Table 1, the same number of variations of the macro-model were modelled. Fig.7 gives the von Mises stress distribution at the crack tip of a macro-model. As already mentioned in Chapter 2, the geometry itself with the structural pre-crack stays the same for all variations, however, the used seam length varies. By evaluating ΔK for each crack length and putting the results into a common diagram gives the expected linear relationship (see region II in Fig.2) between ΔK and the crack length a (Fig.8).

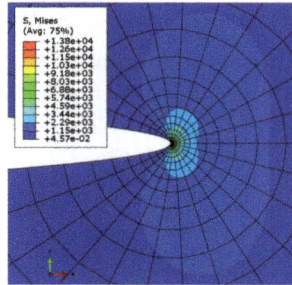

*Figure 7. Stress field at the crack tip of the macro-model
that is used to calculate the ΔK values.*

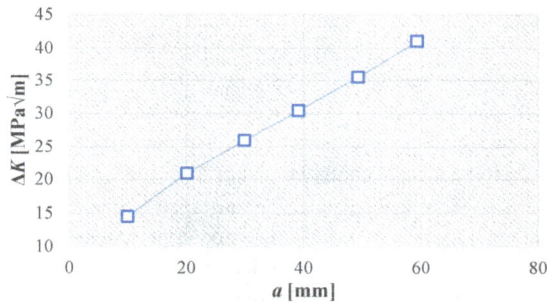

Figure 8. ΔK in dependence of the crack length.

In the submodel, the crack can form in a grain on different slip bands depending on the grain orientation, i.e. on the stress field inside and in the vicinity of the grain – generally, the stress field in the microstructural model (Fig.9-12) is influenced by boundary conditions (displacements from the global model), microstructural configuration (i.e. orientation and shape of crystals), elastic constants and geometrical factors (notch, formed cracks etc.). The slip bands of a certain grain have the same orientation and are distanced one from each other with an offset. Furthermore, each slip band is divided in four segments meaning that a crack does not form instantaneously through the whole grains but it forms in a segmental manner. The condition for the formation of crack segments is contained in the Tanaka-Mura (Eq. 2) and it reads as follows: the average shear stress $\Delta\bar{\tau}$ on a slip band segment needs to exceed two times the *CRSS*. Additionally, the Eq. 2 gives the number of cycles dN (i.e. N_s) that are spent for every formed crack segment. Furthermore, after the crack condition has been satisfied in one step, the model gets updated with the latest crack and remeshed in the following step where the condition is applied again and a new weakest crack segment is traced.

The von Mises stress state before the first formed crack segment is shown in Fig.9. Figures from 10 to 12 show the stress states after 5, 15 and 25 broken crack segments. Those four figures show the simulation results for the crack length of 20.1 mm. Results for other considered structural cracks from table 1 are principally similar and, therefore, not shown in the paper.

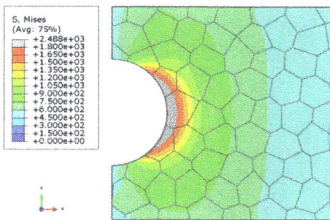

Figure 9. Von Mises stress distribution in the undamaged microstructural model(a=20.1mm).

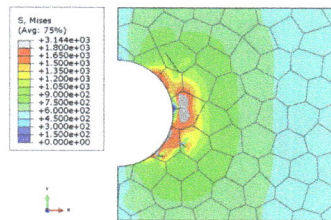

Figure 10. Microstructural model with 5 broken crack segments.

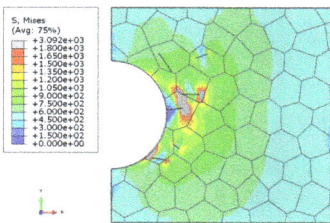

Figure 11. Microstructural model with 15 broken crack segments.

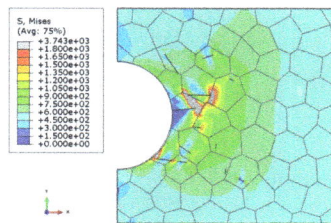

Figure 12. Microstructural model with 25 broken crack segments.

Materials Research Forum LLC
https://doi.org/10.21741/9781644901656-1

Generally what happens for all considered structural cracks is that the formation of cracks in the submodel ceases after a certain number of steps – the number varies from one model setup to another. The reason for this is that grains that are favourable for cracking on the basis of the aforementioned condition fade out.

In order to determine the fatigue crack growth rate da/dN, both the cycles dN and the accompanying length da of each individual crack segment that formed in the microstructural model have to be determined. The crack length da can easily be quantified using the ABAQUS graphical interface or can be rather gathered from output data.

By taking the measured da and the correlated dN one can evaluate the growth rate da/dN for each step. It is necessary to indicate that the common growth rate da/dN is not a constant value, but it fluctuates as it can be seen in Fig.13, for the case of the 20.1 mm long structural crack. It can be noticed that Fig.12 contains 25 broken segments while only 11 are considered in Fig.13, where da/dN is plotted. An explanation for this is that starting from step 12 in Fig.13, the number of required cycles for further formed cracks becomes significantly higher, meaning directly that the da/dN drops down to a negligible value. This leads to the assumption and consequence that the crack growth rate should be neglected in this relatively advanced stage.

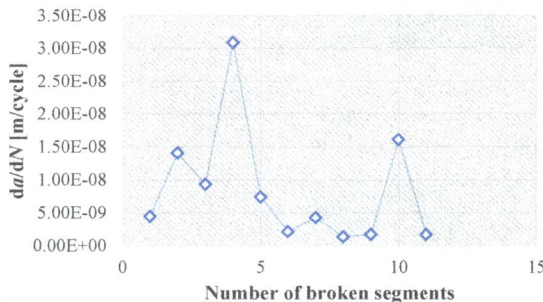

Figure 13. Crack growth rate in relation to the number of broken segments for the extension of the 20.1 mm long structural crack.

In order to get a typical da/dN versus ΔK plot as in Fig.2, the fluctuating crack growth rate from, e.g. Fig.13, needs to be averaged. For the structural crack length of 20.1 mm this results in da/dN = 8.49421x10^{-9}, which seems to be appropriate according to literature [3,

19]. This averaged growth rate can be considered as extensional growth rate for the structural crack. Similarly, for the remaining structural crack lengths. Finally, the results for da/dN and for ΔKs (Fig.8) are put into a common log da/dN versus log ΔK diagram, Fig.14. The resulting six single points were interpolated with a straight line from which the material constants of Paris equation (1) were determined, the slope of the line is $m = 2.75$ and the y-axis intercept $C = 3.8 \times 10^{-12}$, which agree quite well with the experimentally determined values of $m = 2.75$ and $C = 1.43 \times 10^{11}$ [3].

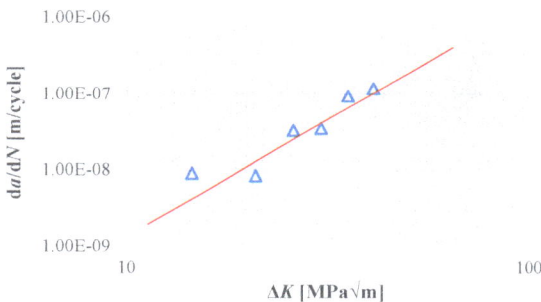

Figure 14. Extensional crack growth rate da/dN of different structural cracks versus ΔK of the same cracks. Data are shown in log-log scale.

4. Discussion and conclusions

The aim of this chapter was to numerically determine the Paris law material constants C and m and this was successfully accomplished by applying a two-scale model. Namely, stress intensity factors for six different structural cracks were calculated using the macro-model based on classical LEFM. In the second part, the fatigue crack growth rates at the tips of same structural cracks were determined by using the microstructural model enriched by the Tanaka-Mura equation.

Having determined the stress intensity factors and the crack growth rates separately, the values have been visualized afterwards in the common log-log diagram (figure 14). Considering that the Paris law can be approximated by a straight line in the diagram for the case of stable crack growth, the material constants have been determined then as $m = 2.75$ and $C = 3.8 \times 10^{-12}$.

Since this study was the first, to the authors' knowledge, attempt to determine the Paris constants in a numerical way as well as taking into account the microstructure of the material, and considering that the literature values of the investigated material ($m = 2.75$ and $C = 1.43 \times 10^{-11}$) match the numerically obtained ones quite good, the study and its working hypothesis seem to prove their worthiness. It is to be supposed that the results determined here may even become better during further research.

References

[1]   Esslinger V, Kieselbach R, Koller R and Weisse B, 2004. Eng. Fail. Anal. 11(4) 515-35. https://doi.org/10.1016/j.engfailanal.2003.11.001

[2]   Paris P and Erdogan F, 1963. J. Basic Eng. 85(4) 528-33. https://doi.org/10.1115/1.3656900

[3]   Božić Ž, Mlikota M and Schmauder S, 2011, The. Vjesn. 18(3) 459-66

[4]   Broek D, 1988, The Pratical Use of Fracture Mechanics (Dordrecht: Kluwer Academic Publishers)

[5]   Branco R, Antunes F, Ferreira JM and Silva M, 2009, Eng. Fail. Anal. 16 631-38. https://doi.org/10.1016/j.engfailanal.2008.02.004

[6]   Branco R, Antunes F, Costa D, Yang FP and Kuang ZB, 2012. Eng. Fract. Mech. 96 96-106. https://doi.org/10.1016/j.engfracmech.2012.07.009

[7]   Carrascal I, Casado J, Diego S, Lacalle R, Cicero S and Alvarez J, 2014. Polym. Test. 40 39-45. https://doi.org/10.1016/j.polymertesting.2014.08.005

[8]   Chauhan S, Pawar AK, Chattopadhyay J and Dutta BK, 2016 T. Indian I. Metals 69(2) 609-15. https://doi.org/10.1007/s12666-015-0796-1

[9]   Ancona F, Palumbo D, Finis RD, Demelio G and Galietti U, 2016. Eng. Fract. Mech. 163 206-19. https://doi.org/10.1016/j.engfracmech.2016.06.016

[10]   Szata M and Lesiuk G, 2009, Arch. Civ. Mech. Eng. 9(1), 119-34. https://doi.org/10.1016/S1644-9665(12)60045-4

[11]   Tanaka K and Mura T, 1981, J. Appl. Mech. 48(1) 97-103. https://doi.org/10.1115/1.3157599

[12]   Tanaka K and Mura T, 1982, Metall. Trans. A 13(1) 117-23. https://doi.org/10.1007/BF02642422

[13]   Jezernik N, Kramberger J, Lassen T and Glodez S, 2010, Fatigue Fract. Eng. M. 33(11) 714-23

[14]   Glodez S, Jezernik N, Kramberger J and Lassen T, 2010, Adv. Eng. Softw. 41(5) 823-29. https://doi.org/10.1016/j.advengsoft.2010.01.002

[15]   Božić Ž, Schmauder S, Mlikota M and Hummel M, 2014, Fatigue Fract. Eng. M. 37(9) 1043-54. https://doi.org/10.1111/ffe.12189

[16]   Mlikota M, Schmauder S, Božić Ž and Hummel M, 2017, Fatigue Fract. Eng. M. (accepted)

[17]   SIMULIA ABAQUS Documentation

[18]   Roos E and Eisele U ,1988, J. Test. Eval. 16(1) 1-11. https://doi.org/10.1520/JTE11045J

[19]   Božić Ž, Schmauder S and Mlikota M, 2011, Key Eng. Mater. 488-489 525-28. https://doi.org/10.4028/www.scientific.net/KEM.488-489.525

Multiscale Fatigue Modelling of Metals
Materials Research Foundations **114** (2022) 16-36

Materials Research Forum LLC
https://doi.org/10.21741/9781644901656-2

Chapter 2

Calculation of the Wöhler (S-N) Curve Using a Two-Scale Model

M. Mlikota[1], S. Schmauder[1], Ž. Božić[2]

[1] Institute for Materials Testing, Materials Science and Strength of Materials (IMWF), University of Stuttgart, Pfaffenwaldring 32, 70569 Stuttgart, Germany

[2] Faculty of Mechanical Engineering and Naval Architecture, University of Zagreb, I. Lučića 5, 10000 Zagreb, Croatia

Abstract

This chapter deals with the initiation of a short crack and subsequent growth of the long crack in a carbon steel under cyclic loading, concluded with the estimation of the complete lifetime represented by the Wöhler (*S-N*) curve. A micro-model containing the microstructure of the material is generated using the Finite Element Method followed by the calculation of respective non-uniform stress distribution is calculated afterwards. The number of cycles needed for crack initiation is estimated on the basis of the stress distribution in the microstructural model and by applying the physically-based Tanaka-Mura model. The long crack growth is handled using the Paris law. The analysis yields good agreement with experimental results from literature.

Keywords

Multiscale Modelling, Fatigue, Crack Initiation, Lifetime Estimation, Wöhler (*S-N*) Curve

Contents

Originally published in the International Journal of Fatigue, 114 (2018) 289-297
https://doi.org/10.1016/j.ijfatigue.2018.03.018

1.   Introduction

Fatigue is one of the most important modes of failure in many mechanical components since it may occur even if the stresses in the critical regions are below the elastic limit. This kind of failure can be understood by a simple example. If one tries to break a wire by hands, it can be done in two ways. One way is to stretch it, shear it or bend it in one direction what would typically require a lot of effort. On the other hand, if one tries to bend it cyclically by a hogging-sagging method it would take considerable less time and effort to cause the breakage [1]. This fact highlights the specificity and importance of the fatigue process.

As described above, fatigue is characterized by a series of forward and reverse loadings. Over the course of experienced cycles, defects in the form of dislocations multiply and accumulate within the material. This behavior results in an increase in the dislocation density [2-3]. Consequently, in many materials, strain is localized in the form of slip bands or slip lines, which appear fine and sharp in favorably oriented grains (crystals) within the microstructure. As more cycles accumulate, more grains display signs of slip bands, the existing bands widen, and ultimately some of them develop into short cracks [4]. The number of slip systems in metals is usually high, and those that are active possess an orientation near to the planes with maximum shear stress. Under uniaxial loading the planes of cracks are always inclined approximately 45° to the direction of the applied loading. In the course of further cyclic loading, cracks formed along these slip bands grow and link together (Fig. 1).

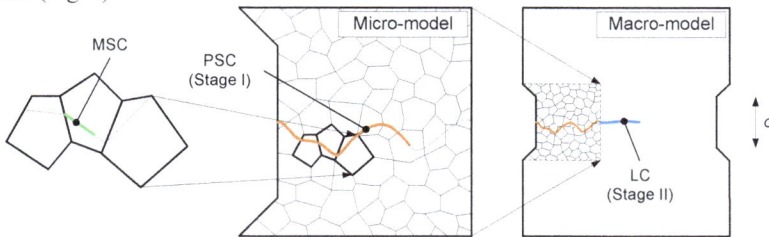

Figure 1.   Stages of fatigue crack during its growth: Left – A microstructurally short crack (MSC) on the grain size scale; Middle – a physically short crack (PSC) on the microstructure size scale (Stage I); and Right – a long crack (LC) on the specimen size scale (Stage II).

Once cracks have nucleated due to strain accumulation, i.e. cyclic slip, they grow in an early stage as microstructurally short cracks (MSC). In this early stage, the MSCs are typically in the order of the material's grain size (Fig. 1 – Left) [5-6]. In metals and alloys they grow predominantly along crystallographic planes at erratic rate because they are highly affected by microstructural barriers such as grain boundaries or other microstructural features [6].

Further, as cracks have grown through several grains they are considered as being physically short crack (PSC) – the length of PSCs is usually in the range from several grains up to 1-2 mm – Fig. 1 – Middle [5]. The crack growth in this stage is predominantly influenced by the microstructure – Fig. 2.

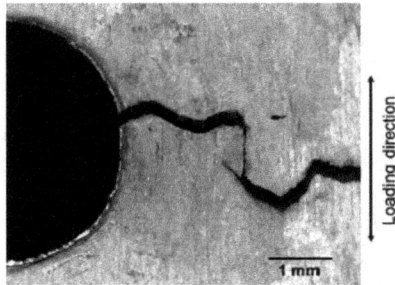

Figure 2.   Crack path influenced by microstructure (Stage I growth) [7].

Upon reaching the end of the PSC regime, the microstructural influences become negligible and such a crack starts propagating in a continuous manner perpendicularly to the outer loading direction, i.e. it develops into a long crack (LC) – Fig. 1 – Right. When the dominant crack has grown to such a size that the remaining ligament can no longer carry the applied load, the component fractures [6]. The change from PSC to LC regime is called the transition from Stage I (crystallographic growth) to Stage II (non-crystallographic growth) or transition from the crack initiation to crack the growth stage [8]. In engineering applications, the first two stages are usually termed as the crack initiation period, while LC growth is the crack growth or propagation period. Those two stages form the complete fatigue lifetime of a specimen or a component. The crack initiation period generally accounts for most of the service life, especially in the high-cycle fatigue (HCF) regime [9], Fig. 3 [10]. Furthermore, by approaching the fatigue limit, the long crack growth contribution diminished.

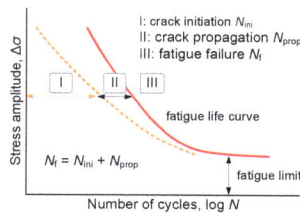

Figure 3.   *Schematic illustration of the two stages of fatigue in ductile metals until failure [10].*

In the LC regime the fatigue crack growth rate (FCGR), da/dN, can be characterized by the stress intensity factor range, ΔK, as a dominant driving force. On the other hand, PSCs usually exhibit a faster growth rate than predicted on the basis of the LC methodology, and they even grow below the threshold of the stress intensity factor, ΔK_{th}, for LC [6,11,12] - Fig. 7 – Right. The fast growth rate of PSCs has often been attributed to the lack of significant crack closure at the early stage of propagation [6,12-14]. In the case of LC growth, crack closure is caused by residual plastic deformations left in the wake of an advancing crack. A crack nucleating at an inclusion particle, a void or a weak grain does not have the prior plastic history to develop closure [14].

Different natures of the crack initiation and propagation stages give rise to the importance of understanding the complete fatigue lifetime and of being able to estimate the lifetime quantitatively. In connection with this, numerical investigation and quantitative evaluation of the fatigue crack initiation and fatigue crack growth processes are of high practical interest.

2.   Fatigue crack initiation and long crack growth calculation

A dislocation model of a double pileup on a single slip band proposed by Tanaka and Mura in 1981 [15-16] is used in this study to numerically describe the crack initiation stage. The number of loading cycles up to fatigue crack initiation is determined by summing the cycles spent for the nucleation of individual micro-cracks that form within the microstructural model of the investigated steel AISI 1141. The number of cycles N_g required for micro-crack nucleation in a single grain can be determined using the physically-based Tanaka-Mura (TM) equation [15-16]:

$$N_g = \frac{8GW_c}{\pi(1-\nu)d(\Delta\bar{\tau}-2CRSS)^2} \qquad (1)$$

Eq. 1 presumes that cracks form along slip bands of grains, depending on the slip band length, d, the average shear stress range on the slip band, $\Delta\bar{\tau}$, the shear modulus, G, the crack initiation energy, W_c, the Poisson's ratio, ν, and the critical resolved shear stress (CRSS) [17].

A very common and often used method for the characterization of the long crack growth is the Paris law, Eq. 1, which gives a relationship between the fatigue crack growth rate, FCGR or da/dN, and the stress intensity factor, $\Delta K(= K_{max} - K_{min})$, at the crack tip during the stable crack growth [18].

A typical fatigue crack growth rate curve – also known as da/dN versus ΔK curve – is shown in Fig. 4. The curve is defined by regions I, II and III.

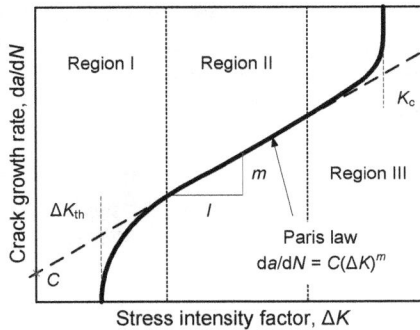

Figure 4.   Typical fatigue crack growth behavior for long cracks [18].

The Paris law relationship can be visualized as a straight line for the region of stable crack growth – region II [19]. The accompanying mathematical equation contains two material parameters C and m, where m represents the slope of the line in Fig. 4 and C the y-axis-intercept [20]:

$$\frac{da}{dN} = C(\Delta K)^m \qquad (2)$$

The stable crack growth rates in the region II are typically in the order of 10^{-9} to 10^{-6} m/cycle. The constants C and m are usually determined as shown in experiments [21-23] and also depend on the material and various influencing factors such as temperature, environmental medium and loading ratio [21,24]. Numerical determination of this constants could also be done [25].

Even though various standardized approaches are available, it is necessary to validate the approaches for the long crack growth characterization. The long crack modelling is well utilized in the field of fatigue research, and accordingly well understood. The methodology is used in this research primarily to accomplish the aim, which is to numerically determine the complete Wöhler (S-N) curve.

The two-scale approach consisting of initiation simulation based on the Tanaka-Mura equation and long crack growth simulation based on Fracture Mechanics has previously been reported [17], [25] and [26]. To differentiate the works published there from the one that follows in Chapters 3 and 4, perhaps it is helpful to repeat the major achievement from the past. In [17], the authors introduced the idea of simulating the fatigue behavior of a material by means of multiscale approach, consisted of Molecular Dynamics based parameter determination, Tanaka-Mura based initiation and Fracture Mechanics based long crack growth. The key aspect of this approach is that the Wöhler curve can be obtained for new materials without need to perform expensive experiments. However, besides of introducing this idea and detailed description of the workflow, the approach has not been applied to any specific case, i.e., no results for numerically obtained Wöhler curve have been published at this time. The aim in [25] was to numerically determine the Paris law material constants C and m and this was successfully accomplished by applying the two-scale model. No results on Wöhler curve determination have been shown. The capability of the crack initiation model based on Tanaka-Mura equation was used in [26] to simulate fatigue crack initiation within the microstructure of medium carbon steel AISI 1141 and to analyse the influence of the overload on the crack initiation process. However, no results on long crack simulation and on Wöhler curve determination have been documented.

3.   Numerical determination of the Wöhler (S-N) curve

In order to achieve the objective of determining the complete Wöhler (S-N) curve of a specimen made out of AISI 1141 steel [27] in this study, a Finite Element Method based two-scale model is applied. The model is split into a macro-model (global model) for the assessment of the global stress field that is transferred to a micro-model (submodel), which was used for the assessment of crack initiation based on the TM equation. Additionally, the macro-model is used to calculate stress intensity factors, ΔK, which are input to the Paris law [18] for the evaluation of the long crack growth.

3.1 Crack initiation modelling

A specimen analyzed in this work is taken from the study of Fatemi et al. [27]. The right image in Fig. 5 shows the stressed three-dimensional (3D) model of the specimen; in this case a half of the specimen due to applied symmetry boundary conditions. The numerical model of the specimen is created and analyzed by using the Finite Element Method (FEM) based software Abaqus. The model is meshed with C3D8R elements, from the ABAQUS element library [28]. The relevant dimensions of the specimen are 141.73 mm height (y-axis) and width of 63.50 mm (x-axis), while the thickness is 2.54 mm (z-axis) and the notch radius 9.128 mm. All other relevant information on the specimen geometry is provided in [27]. The critical site of the specimen – marked with a red square – that becomes vulnerable under cyclic loading is given in Fig. 5 – Right. The specimen failed in experiments from [27] under stress-controlled cyclic loading conditions with the loading ratio $R = 0$, i.e. under fully tensional cyclic loading. The range of applied stress (180–255 MPa) is given in Fig. 10. The model was loaded under the same conditions as the specimen in the experiments [27]. Furthermore, this global model served to provide boundary conditions for microstructural, i.e. crack initiation, analyses. The used submodel (microstructural model; Fig. 5 - Left), located at the notch of the global model, requires the transfer of boundary conditions – in this case displacements – from the global model.

The global model, or macro-model, has been loaded in elastic regime – this especially counts for the notch region where the connection between the global model and the submodel has been established. In this region, the stresses are below the yield stress in all investigated cases, meaning that the transferred displacements from the global model to the microstructural one should not cause stresses at the boundary region between the models that are higher than the yield limit (564 MPa).

The microstructure in the submodel was created on the basis of a typical etched microstructure of AISI 1141 from the study of Mirzazadeh and Plumtree [29]. The average grain size in the model is given as 60 µm, in accordance with the data from the study. There are 253 grains in the model, which are meshed with 171 288 elements in total. Although looking like a 2D model in Fig. 5 – Left, the microstructural model is a 3D deformable shell model meshed with membrane elements with reduced integration (M3D4R). The model is created as a 3D representative volume element (RVE) generated by the Voronoi tessellation technique. According to [28], general membrane elements are often used to represent thin stiffening components in solid structures, such as a reinforcing layer in a continuum. In this study, they are similarly used to model a layer of material inside a bulk, taking additionally the microstructure of the bulk into account. On the other hand, the global model is 3D model with thickness in z-direction of 2.54 mm. Furthermore, according to Abaqus documentation, general membrane elements should be used in 3D models in which the deformation of the structure can evolve in three dimensions, what is the case in the present investigation. Additionally, the software uses a membrane section definition to define the section properties, including the thickness. The thickness can be defined as constant or as spatially varying thickness for membranes using a distribution.

In general, the submodelling technique can be used to drive a local part (submodel) of the model (global model) by nodal results, such as displacements (Node-based submodeling), or by the element stress results (Surface-based submodeling) from the global mesh [28]. The reason that the microstructural shell model is driven in this study by the displacements of the global solid model is that the applied FEM based software Abaqus allows application of the stress-based submodelling only when solid-to-solid models are combined.

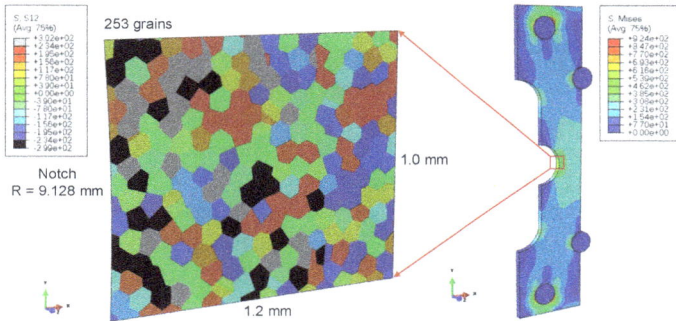

Figure 5.   Left – Shear stresses in a 3D deformable shell submodel, where loading is accomplished by applying stress distributions from the 3D global model to the boundary edges of the submodel; Right – 3D global model of the notched tensile specimen (half).

Concerning the material model definition, an elastic orthotropic material behavior is assumed in the micro-model while pure isotropic elasticity (Young's modulus $E = 200$ GPa, shear modulus $G = 78125$ MPa and Poisson's ratio $\nu = 0.3$. [27]) is adopted in the macro-model. The components of the material stiffness matrix in elastic orthotropic description, i.e. the material elastic constants for cubic crystal symmetry applied in the micro-model are: $C_{11} = C_{22} = C_{33} = 255\,682$ MPa, $C_{12} = C_{13} = C_{23} = 99\,432$ MPa, $C_{44} = C_{55} = C_{66} = 78\,125$ MPa. The constants are calculated using the following equations:

$$C_{11} = \frac{E(1 - v)}{(1 - v - v^2)} \tag{3}$$

$$C_{12} = \frac{E(1 - v)}{(1 - v - 2v^2)} \tag{4}$$

$$C_{44} = G \tag{5}$$

The material parameters of the Tanaka-Mura model, Eq. (1), used in this study are $W_c = 19.0$ kJ/m^2 [30] and the CRSS = 117 MPa [17,31].

Fig. 5 – Left gives the shear stress distribution obtained in the FE analysis for the 256 MPa alternating loading conditions, with an evident influence of the microstructure. Gray and black grains are those where the condition for crack nucleation is satisfied according to the Tanaka-Mura model, which says that the absolute value of average shear stress on a slip band has to be higher than two times the CRSS, i.e. 234 MPa. It is noticeable that these grains, which are conditioned for cracking, are located near the notch.

3.2   Crack growth modelling

In order to characterize the long crack growth numerically, the macro-model for the determination of the stress intensity factor, ΔK, requires certain adjustments at the region of interest, i.e. at the crack-affected region. More specifically, the macro-model requires the usage of a special ABAQUS technique to represent the crack, namely the seam crack. A seam defines in a model an edge or a face that is originally closed but can open as a crack during an analysis [28].

As already mentioned only one half of the specimen needs to be modelled, Fig. 6 – Left. The symmetry is realized by using boundary conditions on the vertical y-axis, which fix displacements in x-direction as well as rotations of any kind.

The long crack growth analysis was performed using classical Linear Elastic Fracture Mechanics (LEFM), where ΔK was calculated in all considered loading cases (180-256 MPa) using Abaqus software.

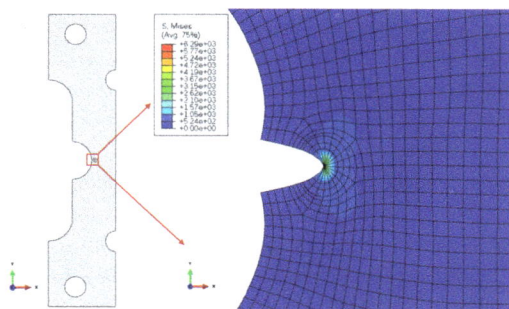

Figure 6. Left – Macro-model for determining the stress intensity factor ΔK; Right – Seam crack and collapsed elements at the crack tip (256 MPa loading case).

The material behaviour is assumed to be purely linear elastic; only a small plastic zone at the crack tip is expected (Fig. 6) thus, no plastic material data are necessary to be used. The isotropic material data for the specimen made of AISI 1141 carbon steel are the same as those that were used in the crack initiation modelling, $E = 200$ GPa, $G = 78125$ MPa and $v = 0$ [27].

For this model, continuum plane stress 8-node elements with reduced integration (CPS8R) were used. Reduced integration has been often used as mean to avoid shear locking in thin shell structures [32], such as the one investigated in this paper. In order to deal with the stress singularity at the crack tip in the case of ΔK calculation, special elements have to be used that are able handling the infinite stresses. This is done in ABAQUS by collapsing one side of an 8-node isoparametric element so that all three nodes from that side have the same geometric location (on the crack tip) [28], as can be seen in Fig. 6 - Right. The crack tip is modelled with a ring of collapsed quadrilateral elements.

4.   Results

A micro-model with a corresponding microstructure selected for the numerical analysis of the fatigue crack initiation lifetime of the specimen made out of AISI 1141 steel is shown in Fig. 5 – Left. The model was loaded with 4 different stress amplitudes, 180, 200, 224 and 256 MPa, in accordance with the data available from the experimental study, Fig. 10.

In the microstructural model consisting of aggregate of grains, the crack can form on different slip bands in each grain depending on the stress field inside and in the vicinity of the grain. Generally, the stress field in the microstructural model (Fig. 5 - Left) is

Multiscale Fatigue Modelling of Metals   Materials Research Forum LLC
Materials Research Foundations **114** (2022) 16-36   https://doi.org/10.21741/9781644901656-2

influenced by loading boundary conditions (own and/or transferred from global model), microstructural configuration (orientation and shape of crystals), material properties (e.g. elastic constants) and geometrical factors (e.g. presence of a notch and/or voids in the microstructure, already formed cracks etc.). The slip bands of a certain grain have the same orientation and are equi-distanced one from each other with an offset, as shown in [33]. Furthermore, each slip band is divided into segments meaning that in each simulated sequence solely one slip line segment of a particular grain gets cracked, i.e. if a segment belonging to one grain breaks in one sequence, it can happen that in the succeeding sequence a segment belonging to another grain breaks [25,26]. Accordingly, the Tanaka-Mura equation, Eq. 1, was adjusted by replacing the grain-breaking cycles N_g with segment-breaking cycles N_s and slip line length d with slip line segment length d_s [33]. The average shear stress range on the slip line segment, $\Delta \bar{\tau}_s$, is an input to the modified Eq. 1, too. In the present investigation, each slip line of each grain is divided into four equally sized segments.

The grains with highest stresses, i.e. the weakest grains in which first MSCs are expected to nucleate, are identified using the Python language based Abaqus plug-in enhanced with the Tanaka-Mura model. The criterion for the identification says that the absolute value of average segmental shear stress, $\Delta \bar{\tau}_s$, has to be higher than two times the CRSS, i.e. 234 MPa, according to the TM equation – Eq. 1. Namely, the denominator of the equation contains part ($\Delta \bar{\tau} - 2CRSS$), which is the indirect explanation for the criterion. It is visible from Fig. 5 – Left, which presents the undamaged micro-model, that the shear stresses differ from grain to grain as well as inside each individual grain. Those grains, identified as vulnerable, are marked with gray and black color in Fig. 5 – Left. Based on that, the first micro-crack is nucleated along the grain slip line segment with the shortest lifetime estimated using the Tanaka-Mura equation, Eq. 1. This means in particular that there can be more segments in the model that fulfilled the stress criterion ($\Delta \bar{\tau} > 2CRSS$), however, the sequence of breaking depends on sequence of fulfilling the condition with respect to the lowest number of cycles for the formation of cracks. The average shear stress range, $\Delta \bar{\tau}_s$, on the slip line segment is an input to the Eq. 1, from the FEM analysis and accordingly has one of the key roles in the simulation of the fatigue crack initiation process.

Upon nucleation of a first segmental crack in the model, the cycles required to nucleate a new crack are again calculated for all grains and all slip line segments in the microstructure on the basis of the new stress field that is locally influenced by the newly nucleated crack. The new stress field can also change the likelihood of some grains for the crack formation, in both directions. In the same manner as in the case of the undamaged microstructure, the next segment that is stressed beyond 2CRSS and that needs a minimal number of cycles for crack nucleation is identified and a crack is introduced in the RVE. Each MSC is formed

Multiscale Fatigue Modelling of Metals   Materials Research Forum LLC
Materials Research Foundations **114** (2022) 16-36   https://doi.org/10.21741/9781644901656-2

in the model in a separate simulation sequence. After the crack condition has been satisfied in one simulation sequence, the model gets updated with the latest crack and remeshed for the following sequence where the condition is applied again and a new weakest slip line segment is traced. The crack generation and remeshing process are done automatically by Python based Abaqus plug-in in every single sequence – by entering the Interaction (Seam Crack option) and Mesh module (Mesh option) of Abaqus software. The simulation sequences run automatically one after the other, too.

Every MSC or segmental crack that formed in the micro-model on the basis of the Tanaka-Mura equation possesses its length, da, and its formation lifetime, dN. By dividing those two output values, one can easily derive the crack growth rate, da/dN. Such a crack growth rate curve plotted as a function of the number of broken slip line segments can be seen in Fig. 7 – Left. It was observed in the present study as well as in a previous one [1] that the crack growth rate develops in an oscillating manner and that it drops down after a certain time. Fig. 7 – Left shows such a behavior in the analysis at 180 MPa loading level. Furthermore, a rough stabilization of the crack growth rate was observed after the drop in the majority of investigations. A similar descending behavior for the short crack growth was reported by Newman et al. [14] and illustrated in Fig. 8 – Right; the dashed lines representing the short crack (PSC) growth and lying to the left of the LC curve (full line) give quite high growth rates at ΔK values less than the LC threshold, ΔK_{th}. For higher loading levels (S2 and S3), as the crack length increases, the data points of the short cracks drop down and eventually approach the LC curve, being coincident with it [14]. In the case of a lower loading (S1), the short crack may even stop growing. This appearance was used to estimate when the crack initiation process finishes; namely, as soon as the crack growth rate drops significantly (Fig. 7 – Left) it was assumed the crack exits in the initiation stage and transits to the long crack growth regime. Accordingly, the number of cycles for the initiation can be estimated by summing the cycles required for all segmental cracks that nucleated until the observed rate drop.

Fig. 8 shows results of the crack initiation analysis for the 256 MPa loading level, i.e. the moment when the initiation stage is considered as completed in terms of required cycles. The left image in Fig. 8 depicts the micro-model containing nucleated cracks, which can be easily perceived with the help of the grain boundaries and the von Mises stress distribution field. The image on the right (Fig. 8) presents the accompanying crack growth rate and the cycles for the initiation completion that were estimated on the basis of the aforementioned assumption that the short crack initiation finishes with the drop of the crack growth rate (red cross), according to Newman et al. [14]. In this case the number of cycles is equal to 29 116 (see Table 1). However, development of cracks proceeded further, what can be realized by the lightly fluctuating curve on the right hand side from the red cross in

Fig. 8 – Right. The number of cycles for the complete crack initiation, N_{ini}, equals the sum of cycles required for all segmental cracks that nucleated until the observed rate drop.

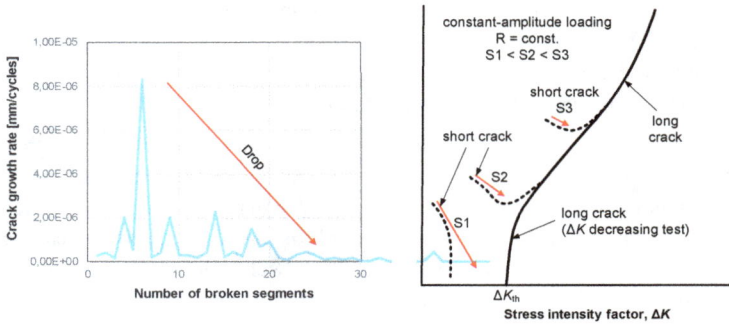

Figure 7.   Left – Fatigue crack growth rate from simulations; Right – Fatigue crack growth rates from experiments for short and long cracks, respectively [14].

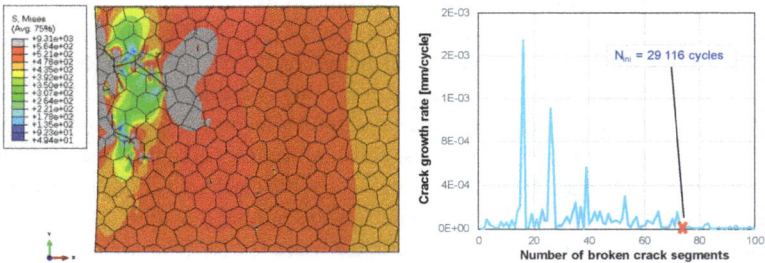

Figure 8. Left – Damaged microstructural model (256 MPa) at the end of the crack initiation stage; Right – Accompanying fatigue crack growth rate.

Concerning the damage evolution, cracks tend to nucleate in the model (Fig. 8 – Left) in a scattered manner and in those grains that are favorably oriented. The cracks occur in the grains where the conditions for micro-crack nucleation according to the Tanaka-Mura equation are fulfilled. Already nucleated crack segments tend to elongate along the whole grain, causing local stress relaxation as well as concentrations at their tips and by that amplifying the likelihood for new crack formation in the vicinity.

By means of Abaqus visualization method, it has been noticed that stresses in the microstructural model, even though purely elastic, go beyond the theoretical yield limit of the material (564 MPa). This happens due to microstructural effects such as grain boundaries, grain shapes and orientations and due to formed cracks whose tips act as stress concentrators. The gray-colored regions in Fig. 8 – Left visualize stresses that surpass the yield limit of the investigated AISI 1141 carbon steel. Even though visualized as "plastic", those regions behave in purely elastic regime, which is defined by the elastic constants for cubic crystal, as described earlier. The gray-colored regions are created by user-defined interval values for output variable (von Mises stress) in Abaqus/Visualization module.

The analysis was performed for different loading levels, given in Fig. 10, and furthermore with 2 different microstructures for each loading case. Both microstructures were generated by the Voronoi technique process and were assigned with the same material properties. The results for crack initiation for all such cases are tabulated in Table 1.

Table 1. Resulting crack initiation cycles for different loading levels.

Stress level [MPa]	180	200	224	256
Cycles	259 580	117 901	64 935	22 346
	369 179	141 613	60 262	24 890
	179 197	77 004	73 063	29 116
	220 186			
Average cycles	257 032	112 173	66 087	25 451

It is assumed in all investigated cases, as exemplarily shown in Fig. 8, that the damage in the microstructural model after the initiation completion is large enough to reach the transition to the long crack growth regime. The initiation number of cycles, N_{ini}, for all loading levels were derived by averaging the values from Table 1, what resulted e.g. for the level of 256 MPa in 25 451 cycles. These cycles were prescribed as a starting number in the long crack growth analysis.

As described earlier, the long crack growth analysis was performed in the next step using the LEFM parameter ΔK, which was calculated using the seam crack and contour integral [28]. The crack size where final failure occurs was estimated by integrating the Paris law (Eq. 2), which is based on ΔK values, and in combination with the critical value of ΔK or fracture toughness, K_{Ic}, that is equal to 67 MPa√m [34]. According to Table 2, the modelled specimen loaded with 180 MPa failed at the crack length of 7.0 mm, where ΔK equals K_{Ic} ($\Delta K = K_{Ic}$, *marked in bold in the table*). Table 2 contains all ΔK values calculated for different crack lengths under 180 MPa loading stress. The same procedure was performed for other loads.

Table 2: Calculated ΔK values for different crack lengths.

Crack length, (mm)	1.5	2.0	3.0	4.0	5.0	6.0	**7.0**	8.0	9.0	10.0
ΔK, (MPa√m)	36.47	39.82	44.97	49.45	54.16	59.66	**66.58**	76.06	90.72	118.47

Since K_{Ic} was reached at a crack length of 7 mm, according to Table 2, the number of cycles to propagate the crack up to failure, N_f, can be easily obtained. This is done by drawing a vertical line from the point where the a-N curve (colored in blue) which is obtained by integrating the Paris law – intersects the magnitude of the crack length a of 7mm (Fig. 9). Besides of that, Fig. 9 depicts the lifetime results of both simulations, i.e. of short crack initiation and long crack growth, including the overall lifetime. Paris constants used here are $m = 3.57$ and $C = 1.00 \times 10^{-12}$ [35].

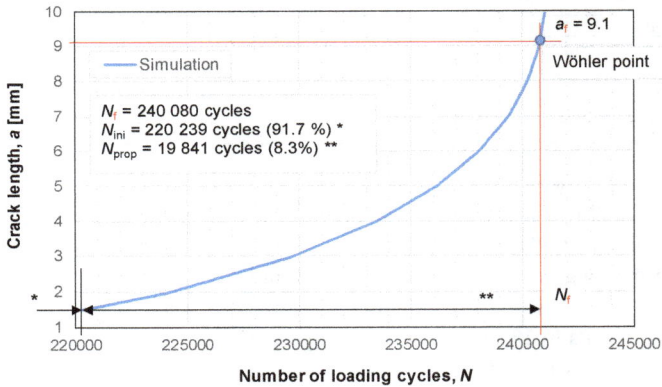

Figure 9. Dependence of the life cycles on the crack length for the 180 MPa loading case.

The initiation number of cycles, N_{ini}, for the 180 MPa loading case were derived by averaging the four values from Table 1, what resulted in 220 239 cycles. These cycles were prescribed as a starting number for the long crack growth. On the other hand, by applying the line intersection method described above, it can be seen that final failure occurs after 240 080 cycles. Furthermore, the number of cycles required to propagate the long crack,

N_{prop}, was calculated by subtracting the N_{ini} (220 239 cycles) from the complete lifetime, N_f, giving N_{prop} equal to 19 841 cycles.

The same procedure was executed for other loading levels. Finally, the complete results (N_f) for all loading cases are presented in Fig. 10 in the form of an Wöhler (S-N) curve and, at the same place, compared with the experimental results. The variance of the crack initiation cycles from Table. 1 is introduced by means of error bars (green horizontal lines).

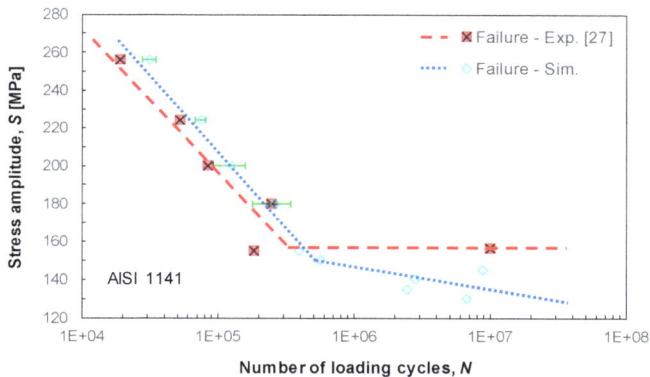

Figure 10. Comparison between simulation and experimental lifetime results [27].

Besides on the slope of the Wöhler curve, the modelling approach allows to determine the fatigue limit, too. The fatigue limit can be derived by just calculating the initiation number of cycles, according to Mughrabi et al. [10] – Fig. 3. The loading amplitude was decreased incrementally, starting from 150 MPa, until the point where there were no cracks appearing inside the microstructural model or where extreme values (> 2E+06 cycles) for the initiation are reached. Already at the loading of around 150 MPa, a noticeable change of the slope was observed. Accordingly, the fatigue limit was chosen to be that one. The results are also given in Fig. 10.

5. Discussions and conclusion

In the elastic region, the relationship between stress and strain remains linear. When a load cycle is applied and removed, the material returns to its original shape and/or length. This conditions are also present in the High Cycle Fatigue, where series of cycles are applied. Nevertheless, the significance of fatigue process lies in the fact that a high number of stress cycles, at a low amplitude - lower than the elastic limit - can cause the part to fail. With respect to that, it is sufficient to use linear elastic models for damage analysis (e.g. LEFM for long crack growth) but also elastically based constitutive laws, in order to obtain satisfactory results. However, the plasticity could be taken into account with respect to constitutive definition in order to sharpen the results and get even better understanding of the process of fatigue. This especially concerns the very small local plastic regions, which in reality form in front of micro-cracks that nucleate in the microstructure. This aspect is one of the prime tasks that need to be tackled in the future work related to the Tanaka-Mura based modelling. However, not a significant influence of this local micro-crack tip plasticity on end result is expected due to negligible crack closure presence at this stage of crack growth, as aforementioned in Chapter 1. On the other hand, it is important to mention that the Tanaka-Mura model itself can be considered (intrinsically) as a plasticity based damage model due to the fact that the dislocations build up has been explicitly considered in its formulation.

The two-scale approach has already been reported in [17], [25] and [26]. However, this is the first time that the approach is applied to numerically construct the Wöhler (*S-N*) curve in its whole extend, including the fatigue limit. Especially, the fatigue limit is a new aspect brought in the research with respect to the quoted papers. Furthermore, a crack initiation criterion based on the crack growth rate drop is a next novelty introduced here.

To conclude, fatigue crack initiation and subsequent long crack growth in a polycrystalline material was successfully simulated using a two-scale fatigue model. The TM equation was applied in the modelling of the crack initiation stage while the classical LEFM was used for the long crack growth, up to the final failure. By combining those two approaches, it was possible to construct the Wöhler (*S-N*) curve numerically. Good agreement was achieved when numerical and experimental Wöhler curves were compared for the case of investigated specimen made out of AISI 1141 steel. It is expected that even better agreement with the experimental results can be achieved by selecting other values for the Tanaka-Mura model parameters, especially the value of CRSS. The aim is to capture this in a future modelling analysis where a parametric study on the influence of these parameters would be performed. The current analysis leads to the following conclusions:

- The process of fatigue crack initiation can be described very well by combining the physically-based TM model with a random grain structure.

- According to Fig. 10, there is a slight shift to the right of the numerically obtained Wöhler curve.

- A potential reason for the slight overestimation of the lifetime curve could be the CRSS value (117 MPa), which was probably overestimated for the AISI 1141 steel in this study. According to the Tanaka-Mura model, the higher the CRSS, the better the durability of a material; or in other words, with higher value of the CRSS, the Wöhler curve shifts to the right.

The 3D shell model based approach that is applied to simulate crack initiation inside the microstructure of the AISI 1141 steel - even though computationally very expensive - leaves a lot of space for further development and numerous future studies; and finally yet importantly delivers satisfactory results in its present state. Nevertheless, a full 3D modelling is one of the important aspects that are planned to be tackled by the authors in future research.

References

[1]   Mlikota M, Schmauder S, 2017, 'Numerical determination of component Wöhler curve', DVM Bericht / Anwendungsspezifische Werkstoffgesetze für die Bauteilsimulation 1684, 111-124.

[2]   Sangid M. D, 2013, 'The physics of fatigue crack initiation', International Journal of Fatigue 57(0), 58-72. https://doi.org/10.1016/j.ijfatigue.2012.10.009

[3]   Polak J, Man, J, 2014, 'Fatigue crack initiation - The role of point defects', International Journal of Fatigue 65(0), 18-27. https://doi.org/10.1016/j.ijfatigue.2013.10.016

[4]   Ewing J. A, Humfrey J. C. W, 1903, 'The fracture of metals under repeated alternations of stress', Philosophical Transactions of the Royal Society of London. Series A, Containing Papers of a Mathematical or Physical Character 200(321-330), 241-250. https://doi.org/10.1098/rsta.1903.0006

[5]   Santus C, Taylor D, 2009, 'Physically short crack propagation in metals during high cycle fatigue', International Journal of Fatigue 31(8-9), 1356-1365. https://doi.org/10.1016/j.ijfatigue.2009.03.002

[6]   Kujawski D, 2001, 'Correlation of long- and physically short-cracks growth in aluminum alloys', Engineering Fracture Mechanics 68(12), 1357-1369. https://doi.org/10.1016/S0013-7944(01)00029-7

[7]   Lorenzino P, Navarro A, 2015, 'Growth of very long "short cracks" initiated at holes', International Journal of Fatigue 71(Supplement C), 64 - 74. https://doi.org/10.1016/j.ijfatigue.2014.03.023

[8]   Klesnil M, Lukas P, 1980, 'Fatigue of metallic materials', Elsevier Scientific.

[9]   Glodez S, Jezernik N, Kramberger J, Lassen T, 2010, 'Numerical modelling of
       fatigue crack initiation of martensitic steel', Advances in Engineering Software,
       41(5), 823 -829. https://doi.org/10.1016/j.advengsoft.2010.01.002

[10]   Mughrabi H, 2015, 'Microstructural mechanisms of cyclic deformation, fatigue
       crack initiation and early crack growth', Philosophical Transactions of the Royal
       Society of London A: Mathematical, Physical and Engineering Sciences
       373(2038). https://doi.org/10.1098/rsta.2014.0132

[11]   Suresh S, Ritchie R O, 1984, 'Propagation of short fatigue cracks', International
       Metals Reviews 29(1), 445-475. https://doi.org/10.1179/imtr.1984.29.1.445

[12]   McEvily A.J, 1989, 'On the growth of small/short fatigue cracks', JSME
       International Journal 32(2), 181-191. https://doi.org/10.1299/jsmea1988.32.2_181

[13]   Christman T, Suresh S, 1986, 'Crack initiation under far-field cyclic compression
       and the study of short fatigue cracks', Engineering Fracture Mechanics 23(6), 953-
       964. https://doi.org/10.1016/0013-7944(86)90139-6

[14]   Newman J, Phillips E, Swain, M, 1999, 'Fatigue-life prediction methodology using
       small-crack theory', International Journal of Fatigue 21(2), 109-119.
       https://doi.org/10.1016/S0142-1123(98)00058-9

[15]   Tanaka K, Mura T, 1981, 'A dislocation model for fatigue crack initiation', Journal
       of Applied Mechanics 48(1), 97-103. https://doi.org/10.1115/1.3157599

[16]   Tanaka K, Mura T, 1982, 'A theory of fatigue crack initiation at inclusions',
       Metallurgical Transactions A 13(1), 117-123. https://doi.org/10.1007/BF02642422

[17]   Bozic Z, Schmauder S, Mlikota M, Hummel M, 2014, 'Multiscale fatigue   crack
       growth modelling for welded stiffened panels', Fatigue & Fracture of Engineering
       Materials & Structures 37(9), 1043-1054. https://doi.org/10.1111/ffe.12189

[18]   Paris P, and Erdogan F, 1963, 'A critical analysis of crack propagation laws',
       Journal of Basic Engineering 85(4), 528-533. https://doi.org/10.1115/1.3656900

[19]   Bozic Z, Mlikota M, & Schmauder S, 2011, 'Application of the K, J and CTOD
       parameters in fatigue crack growth modelling', Technical Gazette 18(3), 459-466.

[20]   Broek D, 1988, The pratical use of fracture mechanics, Kluwer Academic
       Publishers, Dordrecht, The Netherlands.

[21]   Branco R, Antunes F, Ferreira J. M, Silva M, 2009, 'Determination of Paris law
       constants with a reverse engineering technique', Engineering Failure Analysis
       16,631-638. https://doi.org/10.1016/j.engfailanal.2008.02.004

[22] Branco R, Antunes F, Costa D, Yang F. P, & Kuang Z. B, 2012, 'Determination of the Paris law constants in round bars from beach marks on fracture surfaces', Engineering Fracture Mechanics 96, 96-106. https://doi.org/10.1016/j.engfracmech.2012.07.009

[23] Ancona F, Palumbo D, Finis R. D., Demelio G, Galietti U, 2016, 'Automatic procedure for evaluating the Paris Law of martensitic and austenitic stainless steels by means of thermal methods', Engineering Fracture Mechanics 163, 206-219. https://doi.org/10.1016/j.engfracmech.2016.06.016

[24] Szata M, Lesiuk G, 2009, 'Algorithms for the estimation of fatigue crack growth using energy method', Archives of Civil and Mechanical Engineering 9(1), 119-134. https://doi.org/10.1016/S1644-9665(12)60045-4

[25] Mlikota M, Staib S, Schmauder S, and Bozic Z, 2017, 'Numerical determination of Paris law constants for carbon steel using a two-scale model', Journal of Physics: Conference Series 843(1), 012042. https://doi.org/10.1088/1742-6596/843/1/012042

[26] Mlikota M, Schmauder S, Bozic Z, and Hummel M, 2017, 'Modelling of overload effects on fatigue crack initiation in case of carbon steel', Fatigue & Fracture of Engineering Materials & Structures 40(8), 1182-1190. https://doi.org/10.1111/ffe.12598

[27] Fatemi A, Zeng Z, & Plaseied A, 2004, 'Fatigue behavior and life predictions of notched specimens made of QT and forged microalloyed steels', International Journal of Fatigue 26(6), 663-672. https://doi.org/10.1016/j.ijfatigue.2003.10.005

[28] SIMULIA ABAQUS Documentation.

[29] Mirzazadeh M. M, and Plumtree A, 2012, 'High cycle fatigue behavior of shot-peened steels', Metallurgical and Materials Transactions A 43(8), 2777-2784. https://doi.org/10.1007/s11661-011-0830-9

[30] Deimel P, & Sattler E, 1998, 'Non-metallic inclusions and their relation to the J-integral, $J_{i, phys}$, at physical crack initiation for different steels and weld metals', Journal of Materials Science 33(7), 1723-1736. https://doi.org/10.1023/A:1004368213567

[31] Siegfried S, and Immanuel S, 2016, Multiscale materials modeling, approaches to full multiscaling, De Gruyter, Berlin, Boston.

[32] Tsach U, 1981, 'Locking of thin plate/shell elements', International Journal for Numerical Methods in Engineering, 17(4), 633-644. https://doi.org/10.1002/nme.1620170410

[33] Jezernik N, Kramberger J, Lassen T and Glodez S, 2010, 'Numerical modelling of fatigue crack initiation and growth of martensitic steels', Fatigue & Fracture of Engineering Materials & Structures 33(11), 714-723. https://doi.org/10.1111/j.1460-2695.2010.01482.x

[34] Yang L, and Fatemi A, 1996, 'Impact resistance and fracture toughness of vanadium-based microalloyed forging steel in the as-forged and Q&T conditions', Journal of Engineering Materials and Technology 118, 71-79. https://doi.org/10.1115/1.2805936

[35] Hui W, Zhang Y, Zhao X, Xiao N, Hu F, 2016, 'High cycle fatigue behavior of V-microalloyed medium carbon steels: A comparison between bainitic and ferritic-pearlitic microstructures', International Journal of Fatigue 91(1), 232 - 241. https://doi.org/10.1016/j.ijfatigue.2016.06.013

Multiscale Fatigue Modelling of Metals   Materials Research Forum LLC
Materials Research Foundations **114** (2022) 37-65   https://doi.org/10.21741/9781644901656-3

Chapter 3

On the Critical Resolved Shear Stress and its Importance in the Fatigue Performance of Steels and other Metals with Different Crystallographic Structures

M. Mlikota[1], S. Schmauder [1]

[1] Institute for Materials Testing, Materials Science and Strength of Materials (IMWF), University of Stuttgart, Pfaffenwaldring 32, 70569 Stuttgart, Germany

Abstract

This study deals with the numerical estimation of the fatigue life represented in the form of strength-life (S-N, or Wöhler) curves of metals with different crystallographic structures, namely body-centered cubic (BCC) and face-centered cubic (FCC). Their life curves are determined by analyzing the initiation of a short crack under the influence of microstructure and subsequent growth of the long crack, respectively. Micro-models containing microstructures of the materials are set up by using the finite element method (FEM) and are applied in combination with the Tanaka-Mura (TM) equation in order to estimate the number of cycles required for the crack initiation. The long crack growth analysis is conducted using the Paris law. The study shows that the crystallographic structure is not the predominant factor that determines the shape and position of the fatigue life curve in the S-N diagram, but it is rather the material parameter known as the critical resolved shear stress (CRSS). Even though it is an FCC material, the investigated austenitic stainless steel AISI 304 shows an untypically high fatigue limit (208 MPa), which is higher than the fatigue limit of the BCC vanadium-based micro-alloyed forging steel AISI 1141 (152 MPa).

Keywords

Fatigue, Fatigue Life Curves, Numerical Analysis, Microstructure, Crystallographic Structure, Critical Resolved Shear Stress, Fatigue Limit

Originally published in the Journal of Metals (2018), 8, 883
https://doi.org/10.3390/met8110883

Contents

1.   Introduction

The approaches for the fatigue design and analysis are used to estimate when, if ever, a cyclically loaded specimen or machine component will lose its integrity over a period of time due to the fatigue. For the purpose of representing the practical recommendations of such approaches, the strength-life (*S-N*) diagram is often used. The *S-N* diagram provides the bearable stress (or the fatigue strength) *S* versus life cycles *N* of a material. The results are generated typically from tests by using standard laboratory-controlled specimens subjected to a cyclic loading as well as numerical approaches that are becoming more and more relevant recently. The approaches or methods of fatigue failure analysis combine science and engineering [1].

A scientific approach to the question of fatigue strength is to consider the effects of crystal structure on fatigue mechanisms [2–5]. Researchers from the field of fatigue are aware of the ratio between endurance (or fatigue) limit and ultimate tensile strength, σ_e/σ_u. This ratio is also known as the fatigue ratio and is typically higher for ferrous materials (including steels), which are of the body-centered cubic (BCC) type, than for non-ferrous materials, which are face-centered cubic (FCC). Furthermore, ferrous materials generally show a sharp "knee" in the *S-N* diagram at about 10^6 cycles, after which the strength-life curve increasingly flattens. The strength at this point is known as the fatigue limit. Interestingly, most other materials exhibit a gradual flattening between 10^7–10^8 cycles (Fig. 1). Although these effects have been explained by some researchers in terms of strain ageing and dislocation locking [5], there is also evidence that crystal structure plays an important role.

Ferro et al. [2,5] examined two groups of iron-nickel alloys, which included an alloy with 96.5% iron and pure nickel (Fig. 1). In these tests, the ferrous or iron-rich group (BCC) exhibited consistently higher fatigue strengths than the nickel-rich group (FCC), as well as showing a definite fatigue limit at about 10^6 cycles. Fig. 1 illustrates these effects. All materials had the same geometry (More-type specimens), all had the same preparation process (annealing) and all have been tested under same cyclic conditions. More details on the preparation and fatigue experiments can be found in [2,5].

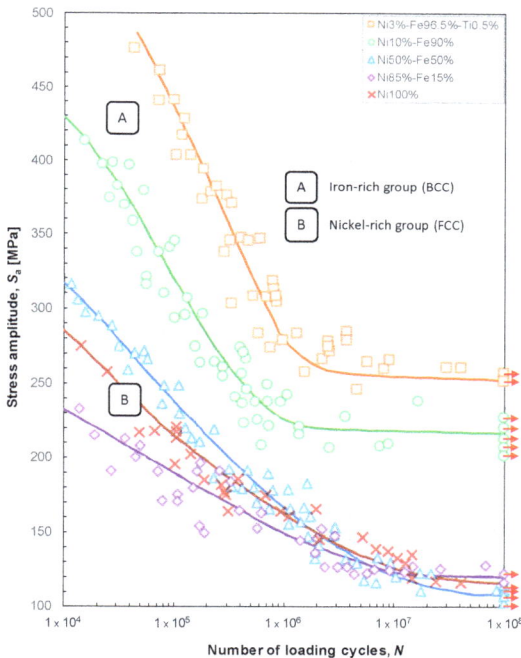

Figure 1. Comparison of fatigue life curves of iron-rich, which are of the body-centered cubic (BCC) type, with nickel-rich alloys, which are face-centered cubic (FCC). Fatigue life curves illustrate the existence of definite fatigue limits in BCC materials, reproduced from [2,5], with permission from Taylor & Francis, 1964.

These same authors collected data on the fatigue ratio, σ_e/σ_u, for a large number of pure materials with different crystal structures and concluded that the ratios of BCC materials were consistently higher than for FCC and HCP (hexagonal close-packed) materials [3]. Table 1, lists fatigue ratios for a number of materials in the various crystal systems. The comparisons are most meaningful for pure metals in which the effects of alloying, aging, etc., are absent.

According to Grosskreutz [5], there is certainly evidence that the crystal structure plays a contributory role in determining the fatigue limit under constant stress amplitude loading. Besides the importance of the crystal structure, the aforementioned differences could be explained in terms of dislocation movement through the crystal. A question that can be posed is why should this influence exist. Since the reasons have not been proven, some general statements can be made. First, coherent slip leading to well-defined bands is not as likely in BCC structures simply because there are so many available slip systems, usually 24 compared to 12 in FCC structures and 3 to 6 in HCP materials, depending on the material (titanium is a special case, with 12 possible slip systems). Therefore, slip is well dispersed in BCC metals and slip band cracking is not as easily achieved. Furthermore, slip activity at a crack tip in BCC metals is not "exhausted" easily by hardening. Therefore, energy is consumed that might otherwise be available for crack extension. This capability to keep well-dispersed dislocation mobility, also in the case of extended fatigue cycling, is seemingly the most significant characteristic of the BCC system with respect to fatigue resistance. A second, assumingly related and equally relevant, reason for the superior performance of the BCC-based systems under fatigue conditions is the larger stress required to move dislocations. This stress is called the critical resolved shear stress (CRSS) and may be 100 times as large in BCC as in FCC structures. The yield stress of BCC metals is correspondingly higher, too. This fact, together with the work hardening rates, which are higher in BCC-based systems, affects the fatigue limit. Namely, BCC metals possess higher fatigue limits than FCC metals, placing them for that reason closer to the ultimate strength.

Accordingly, the effects of crystal structure and the CRSS provide an interesting investigation site, which could eventually provide clues on effective prediction of materials that are more fatigue resistant. Materials Science and numerical methods are promising in combination that possesses capabilities to shed light on this not sufficiently resolved research topic. The aim of this paper is to provide relevant information on the CRSS and to numerically investigate its importance in the fatigue performance of steels and other metals with different crystallographic structures.

Table 1. Fatigue ratios for a number of materials in the various crystal systems, data from [5].

Lattice	Material	σ_u (MPa)	σ_e (MPa)	σ_e/σ_u
BCC	W	1372	834	0.61
	Mo	696	500	0.72
	Ta	308	265	0.86
	Nb	294	225	0.77
	Fe (+0.2% Ti)	265	182	0.69
	Mild steel (0.13% C)	421	224	0.53
	4340 steel	1103	482	0.44
FCC	Ni	303	108	0.36
	Cu	301	110	0.37
	Al	90	34	0.38
	2024-T3 Al	483	138	0.29
	7075-T6 Al	572	159	0.28
HCP	Ti	703	414	0.59
	Co (+0.5% Ti)	521	165	0.32
	Zn	145	26	0.18
	Mg	182	30	0.16

2.   Materials and Methods

2.1   Compression Testing of Small Pillars

A major interest in present-day Materials Science is to understand material deformation and failure mechanisms that are present in a vast number of applications and at different scales (predominantly at micro- and nanoscales). A method based on a new type of compression testing of pillars of the order of nano- and micrometer size has been developed and thereby opened a novel perspective on the investigation and measurement of the CRSS value for various materials. The method is becoming increasingly popular due to its cost-efficient and relatively simple procedure and the ability to analyze deformation mechanisms and material properties by focusing attention on a restricted material volume isolated from single crystals (Fig. 2). The experimental technique used to isolate these small-sized pillars is Focused Ion Beam (FIB) machining. Furthermore, the method of pillar compression testing enables the investigation of specific areas on the surface of a polycrystalline material, and thereby deeper insight into the underlying dislocation mechanisms that eventually contribute to the plastic flow resistance [6].

The pioneering research in the field of pillar testing published by Uchic et al. [7] in 2004 showed the example of cylindrical Ni micropillars, which at that time was an entirely new behavioral regime. Namely, the investigated pillar exhibited low hardening rates and discrete strain bursts (see Figure. 3 for the two effects), and a power-law relationship between CRSS and pillar diameter, or the so-called size effect (see Fig. 4e) [8]. It has been noted that the CRSS values of both FCC and BCC pillars decrease as the pillar size increases, conforming with inverse power-law scaling [7,9,10]. A large amount of publications [7–36] regarding small-scale plasticity are available today.

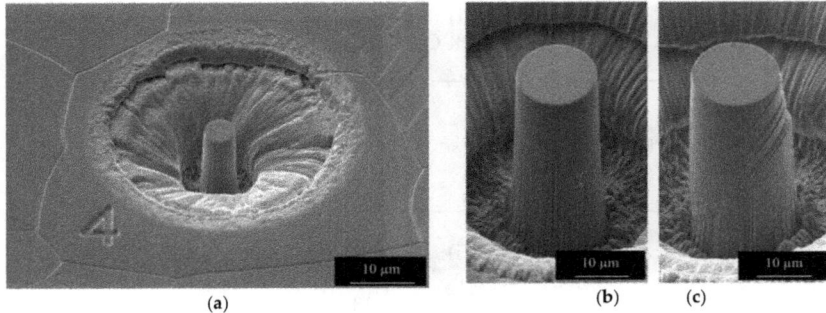

(a) (b) (c)

Figure 2. Example of (a) 5 µm and (b) 15 µm pillars cut within large grains, and (c) deformed large pillar after compression testing, reproduced from [11], with permission from Elsevier, 2013.

Fig. 2 (a, b) show an example of a pillar cut within the austenitic grain (austenitic stainless steel 316L), done by Monnet and Pouchon [11]. After compression, the deformation markings visible as slip traces on the surface of investigated cylindrical single-crystalline pillars indicate single slip mechanism, as Fig. 2c depicts. Fig. 3 gives resolved shear stress vs. strain curves (deduced from the force vs. displacement curves by using Schmid factors) of the pillars considered in the study. The strain bursts that emerge as flat regions have been observed as in the study of Uchic et al. The first strain burst has been understood to be a product of the first motion of a dislocation while the succeeding strain bursts represent collective movement of dislocations in an avalanche-like manner [11–14]. The small-scale yield (shear) stress, i.e. CRSS, for pillars is determined as the stress at which the first strain burst occurs [10].

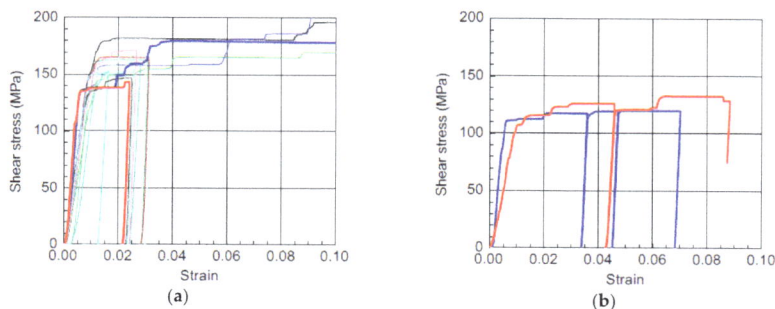

Figure 3. Resolved shear-stress vs. strain curves of (a) 5 μm pillars and (b) 15 μm pillars, reproduced from [11], with permission from Elsevier, 2013.

A significant discrepancy in the value of the CRSS measured for the 5 μm pillars, 160 MPa (Fig. 3a), is to be noted with respect to the larger 15 μm pillar, where the CRSS is equal to 110 MPa (Fig. 3b). This confirms that the size effect is also present in 316L steel. Another observation that is in analogy to the general findings from the pillar compression testing are the fluctuations of the CRSS that decrease as the pillar size increases (Fig. 3) [11].

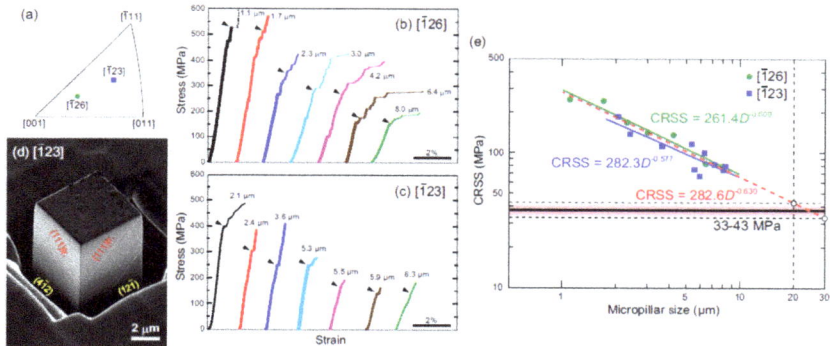

Figure 4. Compression tests of single-crystalline micropillars of the equiatomic CrMnFeCoNi HEA. (b,c) Selected stress-strain curves of single-crystalline micropillars with loading-axis orientations of (b) [$\bar{1}$26], and (c) [$\bar{1}$23], respectively; (d) Secondary-electron image taken in a scanning electron microscope showing {111} slip traces on the side surfaces of a deformed micropillar with [$\bar{1}$23] orientation; (e) Size dependence of critical resolved shear stress (CRSS) for {111} <101> slip, reproduced from [15], with permission from the author, 2016.

Okamoto et al. [15] investigated the behavior of high-enthalpy alloy (HEA) CrMnFeCoNi-based pillars (see Fig. 4) under similar compression tests. Fig. 4 (b,c) show cases where these researchers applied two different loading-axis orientations ($[\bar{1}26]$ and $[\bar{1}23]$, Fig. 4a) in order to obtain stress-strain curves of micropillars. Values of the CRSS are calculated by using the small-scale yield stress magnitudes and the Schmid factors that correspond to the two investigated orientations (0.488 and 0.467 for $[\bar{1}26]$ and $[\bar{1}23]$, respectively) and are given in Fig. 4e versus micropillar size. It follows from Fig. 4e that the CRSS values for the two considered orientations match to each other over the whole range of investigated micropillar sizes. This undoubtedly indicates that the CRSS value for slip on (111)[101] are independent of crystal orientation. This has been used by the authors [15] to combine the data points from the two tests into one master curve, shown in Fig. 4e by the red dashed line. Similar to what has been recognized in single crystalline micropillars of many other FCC and BCC metals [7,9,12,16–18], the size effect has been also observed for the CrMnFeCoNi HEA.

2.2   Size Effect or A Power-Law Relationship between CRSS and Pillar Diameter

A size effect, invoked in previous section, is known to affect two features of plastic deformation: (i) the CRSS becomes stochastic and increases with decreasing pillar dimension; (ii) the hardening rate decreases strongly in pillars of micron scale. However, concerning the CRSS, several studies have shown that the size effect vanishes beyond a given pillar size D_b. In pure nickel, D_b was found being close to 20 μm by Uchic et al. [7] and 30 μm by Dimiduk et al. [19]. In gold, the size effect seems to decrease strongly in pillars of sizes larger than 7 μm [20]. However, in heterogeneous materials, such as Ni-base superalloys, the effect on the CRSS persists even in large pillars [7,11,21]. However, as concluded by various researches [7,8,14,15,17,19,20,22,23], this decrease in CRSS continues until the micropillar size reaches a value of 20 to 30 μm, at which the estimated CRSS may be further taken, for e.g. finite element method (FEM)-based simulations, as the representative value of bulk. In other words, by extrapolating the micropillar data obeying the inverse power-law scaling to the critical pillar size, the bulk CRSS values can be estimated. Accordingly, the bulk CRSS value for the HEA is estimated to between 33 and 43 MPa (Fig. 4e), what fits the range of determined values for pure FCC metals (e.g., ~14 MPa for Al [8]).

The increased strength with reduced size follows a power law that explains the general relationship:

$$\tau_c = KD^{-n} \qquad\qquad (1)$$

where τ_c is the CRSS, D is the top-surface diameter of a cylindrical pillar sample, K is the power-law coefficient and n is the power-law exponent, i.e. the power-law slope [6].

Despite being a matter of major discussions, it is generally agreed that the size effect is a consequence of the dislocation nucleation-governed plasticity, which is apparent from the higher stresses that arise during mechanical deformation of small-sized nanopillars. The explanation of the dislocation nucleation-governed plasticity can be found in its dependency on dislocation storage ability of FCC and BCC systems. Namely, dislocations moving inside the small-sized pillars are attracted to the free surface, and in order to sustain further deformation, new dislocations have to nucleate either inside the pillars or at their surface. This phenomenon especially concerns the strained FCC pillars deep in the sub-micron region where they experience the so-called "hardening by dislocation starvation". In such cases, the pillars remain without dislocations, which vanish from the free surface at faster rates than they multiply inside the bulk [8,20,24]; and to nucleate new dislocations, higher stresses are required. According to experiments on FCC and BCC pillars [18,25–27], the more pronounced size effect in FCC materials, with respect to BCC materials, is related to the lattice resistance to plastic flow (see Section 1). A dislocation-starved condition is unlikely in the BCC system as a longer residence time is attributed to its dislocations, together with the ability to multiply new dislocations before the existing ones exit the pillar surface [27]. On contrary, the longer residence time of dislocations in BCC systems might be an explanation for their higher strength at higher scales, what is also in accordance to the statement from Section 1 about higher values of the CRSS in the BCC systems and their superiority in fatigue performance with respect to the FCC systems.

2.3   Strengthening Mechanisms

Strengthening mechanisms in single-crystals, such as dislocation density (as explained in Section 2.2 in relation to the dislocation residence time) and solute atoms, directly influence the deformation behavior in pillars, and thus the CRSS magnitude. An important strengthening source in industrial materials is the solid solution strengthening, or the alloy friction, that results from solute alloying elements, especially from substitutional elements, within the matrix. Furthermore, a significant content of added elements usually induces formation of precipitates within the matrix of the host material, leading to another strengthening mechanism known as precipitation strengthening. In some industrial materials, like annealed 316L-type austenitic stainless steel, no precipitation is observed, leaving the solid solution strengthening as the prevailing source to the strengthening process [11].

The resistance for the dislocation to move through the crystal, or the CRSS (τ_c), is dictated by the present strengthening mechanisms in the crystal [6]. To recall, the CRSS for compressive failure may have several components: The Peierls-Nabarro stress τ_0, the dislocation hardening τ_{dh}, the solid solution hardening τ_{ssh}, and the strengthening induced by precipitates τ_{ph}. Irrespective if all or just particular strengthening components are present, the linear superposition may be applied when calculating the CRSS (τ_c) [11]:

$$\tau_c = \tau_0 + \left(\tau_{dh} + \tau_{ssh} + \tau_{ph}\right) \tag{2}$$

In the case of high purity materials, with zero dislocation density and with no added alloying elements, the CRSS comprises merely of the Peierls-Nabarro stress τ_0, which is the minimum requirement for initial dislocation motion [28].

Furthermore, the global yield stress R_e comprises of the CRSS and further strengthening mechanisms that are present in bulk metals, namely the grain boundary hardening τ_{gbh} and the phase boundary hardening τ_{pbh}. Eq. 3 summarize all the mechanisms that contribute to the strength of metallic materials.

$$R_e = \tau_c + \tau_{gbh} + \tau_{pbh} = \tau_0 + \tau_{dh} + \tau_{ssh} + \tau_{ph} + \tau_{gbh} + \tau_{pbh} \tag{3}$$

An example of strengthening can be given with iron; The shear strength of large single-crystal samples, in mm size, can go below 10 MPa [6,29,30] at room temperature in the case of high purity iron. Previous studies on iron with different impurity contents [6,31] confirm the importance of interstitial solute atoms in the strengthening of crystals. In general, increased amounts of impurity elements lead to an increased shear strength in the crystal due to more resulting obstacles that hinder the gliding of dislocations. Based on these results, it was estimated that the strengthening contribution of C and O is about 40 MPa; which is significantly higher than the 10 MPa for the high purity single-crystal iron. The strengthening from solute atoms and impurities is the same in different-sized pillars; and it should not be partially responsible for the size effect that has been discussed in Section 2.2 [6].

An interesting study was done by Guo et al. [32] on the measurement of the CRSS for phases in a multiphase material. In this work, coupled with electron backscatter diffraction

(EBSD) technique, micropillar compression was used to evaluate the CRSS of ferrite, which is BCC, and FCC austenite in a cast duplex stainless steel (Z3CN20-09M). Compression tests were carried out by compressing free-standing micropillars of ~5 μm diameter that were fabricated by FIB. The results reveal that BCC ferrite has a much higher strength than FCC austenite; while austenite possesses better ductility than ferrite. The CRSS values are, quantified to be ~194 MPa and ~318 MPa for austenite and ferrite, respectively. Strengthening mechanisms can be regarded as responsible for the higher strength of ferrite. Firstly, solid solution strengthening is introduced by the higher alloying content of substitutional chromium in ferrite (26.74 wt.%) than in austenite (21.11 wt.%) [37]. Although austenite contains a higher nickel content than ferrite, i.e. 9.14 wt.% in austenite compared to 5.18 wt.% in ferrite, it was reported that the solid solution strengthening caused by chromium is greater than that of nickel [38]. Secondly, as already stated in Section 2.2, the low mobility of screw dislocations in BCC crystals usually results in higher strengthening as well as a higher strain hardening rate through dislocation-dislocation interactions or kinetic pileups of the screw dislocation in the area close to the dislocation sources, which in turn leads to an enhanced strength of ferrite phase [25,26,32].

Quantification of the strengthening components is a challenging task but is also necessary for the understanding and eventual modelling of the mechanical behavior of metals. For BCC and FCC materials, the slip event is usually activated in unique crystal systems when the applied stress is higher than their CRSS, which is associated with the intragranular crystal plasticity [32]. Therefore, it is of great importance to be able to determine the CRSS of a phase or more different phases in metallic materials to understand the mechanical behavior of their bulks and to further establish appropriate material and damage models (e.g. Crystal Plasticity and/or Tanaka-Mura (TM) equation/model) for simulation studies.

2.4 Numerical Estimation of Fatigue Behavior with the Help of CRSS

Structural integrity monitoring and characterization of existing damage are of high importance for the service life assessment. An important aspect of the fatigue process is the failure of structures that can occur at load levels accompanied with stresses in the critical regions that are below the material yield stress. This raises the importance of being able to predict the life of these structures before the catastrophic fracture occurs. However, in many cases structures can tolerate cracks up to a certain length, meaning that not every crack is detrimental immediately after its formation. In order to estimate the moment when the crack reaches unstable growth, it is essential to ensure methods being able to describe and quantify the crack growth precisely.

In order to numerically analyse the total fatigue life of structures, i.e. structural components or a specimen, from the moment when the first micro-crack nucleates within a grain up to the moment when the dominant crack evolves into a critical one leading to final failure, a proper multiscale simulation approach is required. Fig. 5 shows a scheme of scales that need to be considered, starting with the nanoscale, going up to micro-/mesoscales and ending up with the macroscale. The up-to-date nanoscale, i.e. atomistic, simulation techniques like ab-initio [39–41] or Molecular Dynamics (MD) [42–44] can provide the relevant material parameters needed at higher length scales of fatigue modelling and simulation scheme. One such parameter is the CRSS on the most active slip plane in a grain, that can be derived from MD simulations, e.g. by using the approach of Hummel et al. [45,46]. Other methods for the derivation of the CRSS are micro-pillar tests, as discussed in previous sections. The CRSS can be used as the input parameter for the micromechanics-based model providing information on the number of loading cycles to nucleate a micro-crack and subsequent growth of a short crack inside the microstructure of the investigated material; or in other words, comprising both the number of cycles to initiate the short crack (TM model [47,48]). These initiation cycles are further transferred to the macroscale fatigue crack growth model based on power law equations (e.g. Paris law), which are finally used to estimate the total fatigue life, up to final fracture. By using the presented modelling workflow, the fatigue of metals can be simulated more or less independently of the experimental input [45,46,49–52].

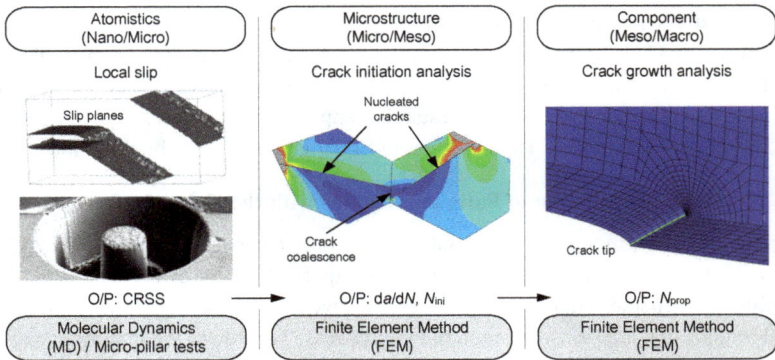

Figure 5. Multiscale approach—Coupling of methodologies at the relevant scales, and accompanying outputs (O/P) (CRSS, da/dN—crack growth rate, N_{ini}—number of stress cycles for crack initiation, N_{prop}—number of stress cycles for crack propagation), reproduced from [34,45,50,53], with permissions from Elsevier, 2016 and John Wiley and Sons, 2014, 2017.

A dislocation model of a double pileup proposed by Tanaka and Mura in 1981 [47,48] is frequently used for fatigue crack initiation analysis (Fig. 5) to determine when a grain, subjected to an outer cyclic loading, will develop a slip band and subsequently a crack [54]. According to their theory of fatigue crack nucleation, the forward and reverse plastic flows within slip bands under cyclic loading are caused by edge dislocations with different signs gliding on two adjacent crystallographic planes. It is assumed that their mobility is irreversible. As reported by the founders of this model, the monotonic build-up of dislocations evolves from the theory of dislocations in a systematic manner. One of the parameters of the model, Eq. 4, is the CRSS on a slip plane. As stated in previous sections, the CRSS is a threshold value of the shear stress along the glide direction that a dislocation needs to surpass in order to start moving. If the resolved shear stress is lower than the CRSS, no dislocation is moving on the glide plane and, consequently, no pile takes place at the grain boundary.

As mentioned above, the number of cycles N_s needed for micro-crack nucleation within a single grain can be derived by means of the physically-based TM equation [47,48]:

$$N_s = \frac{8GW_c}{(1-v)(\Delta\tau_s - 2\text{CRSS})^2\pi d_s} \tag{4}$$

According to TM, micro-cracks form along slip band segments, depending on segmental length d_s, the average shear stress range on the segment $\Delta\tau_s$, the shear modulus G, the crack initiation energy W_c, the Poisson's ratio v, and the CRSS [49,52,55,56].

A more detailed description of the implementation of the TM equation into FEM-based modelling and simulation of the crack initiation process has been reported in publications of the authors of this study [45,46,49–52] and by other researchers, too [55–58]. Some details about mesh, boundary and loading conditions are also given in Section 3.

A well-known and often used method for the quantification of the long crack growth is the Paris law [59], which gives the fatigue crack growth rate, FCGR or da/dN, in relation to the stress intensity factor, ΔK (= $K_{\max} - K_{\min}$), at the crack tip during the stable crack growth. Despite being standard, this well-accepted and proven method is applied in many cases for the characterization of long crack growth. Long crack modelling is used extensively in the fatigue related research and is accordingly well documented and well understood.

The multiscale approach for fatigue simulation, consisting of the CRSS determination, crack initiation simulation based on the TM equation and long crack growth simulation based on Fracture Mechanics, has been previously reported in References [45,46,51,52].

The methodology forms the basis of the presented research work, which has been conducted with the aim to determine numerically the complete *S-N* curve of considered materials.

3. Results

In order to approach the question of fatigue strength in a proper way, it is necessary to consider the effects of crystal structure on fatigue mechanisms. The purpose of this section is to present the derivation of the CRSS parameter for materials of interest by using different methods of determination (micro-pillar tests and MD simulations) and its implementation in numerical analyses of fatigue initiation life, based on the TM model. During the numerical estimation of the fatigue life curves of materials with different magnitudes of the CRSS, an attention has been paid to their crystal structures, too. It has long been known that ferrous materials (BCC) typically show a sharp "knee" (fatigue limit) in the *S-N* diagram at about 10^6 cycles, while most other metals (FCC) exhibit a gradual flattening between $10^7–10^8$ cycles.

Besides the importance of the crystal structure, an assumption that the reason for superior fatigue behavior of BCC metals in general is the higher stress needed to move dislocations, i.e. the CRSS, through their system. Therefore, it is of high interest to analyze numerically the fatigue life curves of materials with different CRSS values, but also by having their crystal structures in view. For this purpose, four materials are selected and are tabulated in Table 2. Besides of the AISI 1141 steel with BCC crystal structure that has been analyzed in [52], three additional materials taken into account are FCC austenitic stainless steel AISI 304, high purity iron (Fe 99.9%) that is BCC and high purity aluminum (Al 1050) with an FCC structure. Their life curves are determined by analyzing the initiation of a short crack and subsequent growth of the long crack separately. Micro-models containing microstructures of the materials are set up by using the FEM and are applied in combination with the Tanaka-Mura equation in order to estimate the number of cycles needed for crack initiation. The long crack growth quantification is accomplished by using the Paris law.

Table 2. CRSS (critical resolved shear stress) values for different materials.

Material	CRSS (MPa)	Method	Source
AISI 304 (X5CrNi18-10)	160	MPT	Monnet and Pouchon [11]
AISI 1141 (40Mn2S12)	117	MD	Hummel et al. [45,46]
Fe 99.9%	35	MPT	Rogne and Thaulow [6]
1050A (Al 99.5%)	14	MPT	Jennings et al. [17]

Values of the CRSS from Table 2 are estimated by using methods presented in Section 2. Namely, the CRSS of 160 MPa for AISI 304 steel is taken from the micro-pillar tests (MPT) performed by Monnet and Pouchon [11]. The 117 MPa high CRSS is taken from the study of Hummel et al. [45,46] performed by using the MD method. This value has been applied for AISI 1141 steel in the numerical determination of fatigue life curves. To quantify the size-strengthening effect and to determine the CRSS of 99.9% pure iron, the experimental data from Reference [6] were plotted and extrapolated, as introduced in Reference [15] (Fig. 4e), as a function of pillar diameter in the double logarithmic plot in Fig. 6. The CRSS is estimated to be 35–45 MPa for this material by applying the aforementioned methodology, explained in Section 2 in detail. Another example is given in Fig. 6 where the data for high purity aluminum from [8] were plotted and extrapolated as a function of pillar diameter. Here the values of CRSS are estimated to be 11–16 MPa. A third example found in literature is for copper [17] with an estimated CRSS of 19–25 MPa. The data for copper are given for comparison reasons, but have not been considered in the numerical study. It is worth noting the observation that the power-law slopes of all three investigated metals (BCC iron and FCC aluminum and copper) are in the same range, namely −0.618 for pure iron, −0.602 for copper and −0.625 for aluminum. Furthermore, the slope of a BCC HEA from the study of Okamoto et al. [15] is given as −0.63.

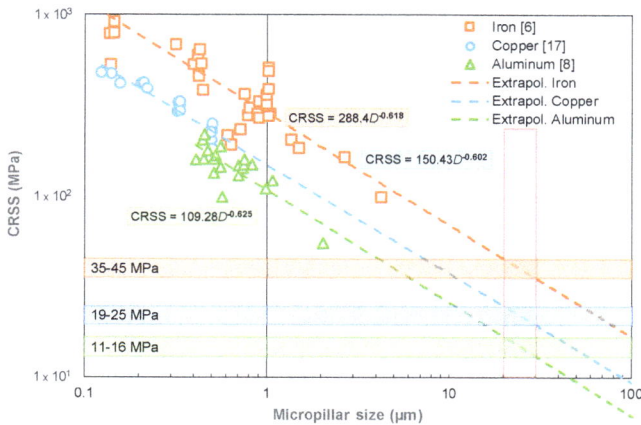

Figure 6. CRSS (critical resolved shear stress) of iron [6], copper [17] and aluminum [8], respectively, as a function of pillar diameter D. Plots of data for different materials show the power law relationship, i.e. the size effect, between pillar diameter and the CRSS, data from [6,8,17].

A notched specimen from the experimental study of Fatemi et al. [60] has been analysed in this work by using the presented multiscale approach for fatigue life prediction (Section 2.4). Fig. 7b contains the three-dimensional (3D) model of the specimen, in a stressed state. The model represents a half of the specimen due to applied symmetry boundary conditions. The FEM-based software ABAQUS has been used to create and analyse the numerical model of the specimen. The model is meshed with 25 752 linear hexahedral elements of type C3D8R, from the ABAQUS element library [53]. Concerning the relevant dimensions, the considered specimen is 141.73 mm high (y-axis), 63.50 mm wide (x-axis), 2.54 mm thick (z-axis) with the notch radius of 9.128 mm. All other details about the specimen are provided in Reference [60]. The spot at the notch visualized by the help of a red square in Fig. 7b becomes critical after putting the specimen in cyclic loading. The specimen fractured, starting from that site, in experiments from Fatemi et al. [60] under stress-controlled cyclic conditions with the loading ratio $R = 0$, i.e. under fully tensional cyclic loading. The ranges of applied stress amplitudes vary from material to material and are given in Fig. 7. Furthermore, this global model serves to provide boundary conditions (in this case displacements) to the microstructural submodel (Fig. 8a), which is located at the notch ground of the global model and is used for the crack initiation analysis.

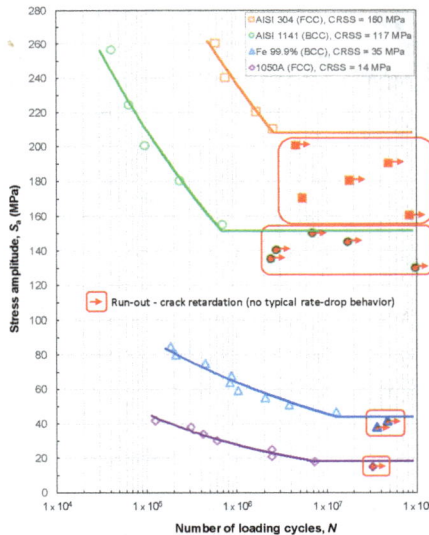

Figure 7. Simulation-based fatigue life curves, which illustrate the existence of definite fatigue limits in BCC and FCC materials depending on the magnitudes of the CRSS.

Numerical models of the 3D global model and of the 3D deformable shell submodel, which have been analyzed in this study for different materials are shown in Fig. 8. In this case, the resulting stresses are given for the models that have material properties of the vanadium-based microalloyed forging steel AISI 1141, and that are loaded with stress amplitude of 256 MPa. Other applied stress amplitudes for this and all other investigated materials are given in Fig. 7. Fig. 8a shows shear stresses in a 3D deformable shell submodel, where loading is accomplished by applying the displacements from the 3D global model to the boundary edges (upper, right and bottom) of the submodel. This specific submodel contains 249 grains and is meshed with 154 708 linear quadrilateral elements of type M3D4R; meaning that each grain contains approximately 621 elements. The microstructures of the investigated materials (Table 3), i.e. their grains, are created by using a Voronoi tessellation technique. The average sizes of grains are given in Table 3. The micro-crack modelling within the microstructural models of these materials has been accomplished by using the TM model, i.e. the model-based criterion that says that the average shear stress on a slip band segment needs to be two times higher than the CRSS of the specific material ($\Delta\tau_s > 2$CRSS). The average shear stresses in the microstructural model (see Fig. 8a) are an input from the FEM-based analysis. The slip band segment that fulfills the criterion and which, next to that, needs the lowest number of cycles to nucleate the crack according to the TM model gets cracked. The microstructural model is remeshed after introducing a newly nucleated crack and the process is repeated until the moment when there are no more segments favorable for cracking. The TM model-based micro-crack modelling considers just the transgranular cracking. Intergranular cracks along the grain boundaries occur in rare situations and only if two already nucleated transgranular cracks are located near the same grain boundary. In such cases, the yield stress is the cracking criterion and no cycles are prescribed to the event.

Fig.8b gives the global model of the notched tensile specimen (half). The same model has been applied for the long crack modelling and simulation by using the Stress Intensity Factors-based Paris Law [59]. The details on this well-established and well-known approach can be found in a previous publication of the authors [52] on the example of AISI 1141. Due to the relative simplicity and general presence of this long crack modelling and simulation approach in the fatigue community, further details are not repeated in this chapter.

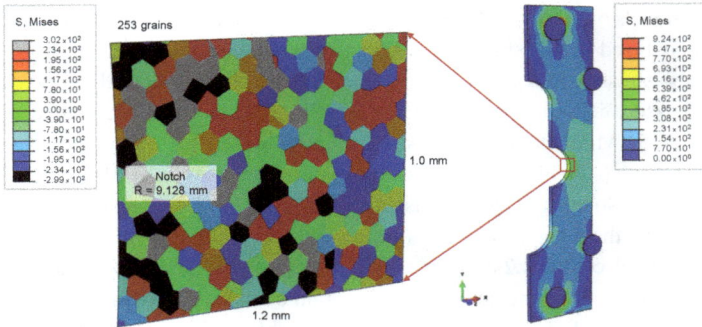

Figure 8. (a) Shear stresses in a 3D deformable shell submodel of the AISI 1141 steel; (b) 3D global model of the notched tensile specimen (half), reproduced from [52], with permission from Elsevier, 2018.

Table 3, contains other simulation-relevant material properties; namely the Young's modulus E, the shear modulus G, the Poisson's ratio v, the three elastic constants C_{11}, C_{12} and C_{44}, the crack initiation energy W_c, and the average grain size d for all materials considered in this study. The length of slip band segment d_s, which is the input to the TM equation, is calculated from the grain size d ($d_s = d/4$) [49,52,55,56]. The constitutive laws of the materials are purely linear elastic, defined by using orthotropic elasticity [53], i.e. the three elastic constants.

Table 3. Mechanical properties of the considered metallic materials.

Material	E (GPa)	G[1] (MPa)	v[1] -	C_{11} (MPa)	C_{12} (MPa)	C_{44} (MPa)	W_c[1] (N/mm)	d[1] (μm)
AISI 304	188	79 000	0.26	233 026	80 820	79 000	69	~30
AISI 1141	200	78 125	0.28	255 682	99 432	78 125	19	~60
Fe 99.9%	205	81 000	0.28	262 073	101 918	81 000	19	~65
1050A	72	26 000	0.33	106 678	52 543	26 000	11	~65

[1] TM model parameters.

Fig.9 shows an example of damaged microstructural model of Fe 99.9% after the analysis has been performed. The model contains visible micro-cracks, which have been introduced by applying the TM equation for the crack nucleation to the FEM-based analysis. A damaged state in the figure represents the end of the crack initiation stage, under a 75 MPa loading stress amplitude.

Materials Research Forum LLC
https://doi.org/10.21741/9781644901656-3

The method for the estimation of the moment when the crack initiation N_{ini} has been accomplished is well-documented in previous publications of the authors of this study [51,52]. An example has been given in Fig. 10 where the fatigue crack growth rate da/dN estimated for the damaged model from Fig. 9 has been plotted in relation to the number of nucleated micro-cracks, i.e. number of broken crack segments. The da/dN can be easily derived by dividing the length, da, of each individual micro-crack that has been nucleated within microstructure and its accompanying nucleation lifetime, dN.

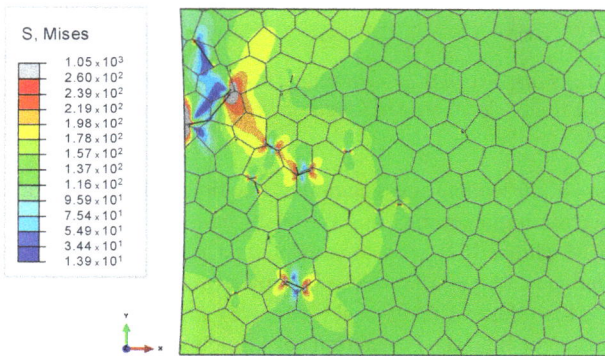

Figure 9. Damaged microstructural model of the Fe 99.9% at the end of the crack initiation stage, under 75 MPa loading stress amplitude.

Namely, it has been observed in the present as well as in previous studies [49–52] that the rate drops down and stabilizes after certain time. Interestingly, Newman et al. in [61] reported a similar declining behaviour during the short crack growth. In such cases, the short crack most likely exits the initiation and enters the long crack growth stage. This appearance, when crack growth enters the regime characterized by significantly low rates (Fig. 10a), is used within the multiscale fatigue simulation to determine cycles related to the crack initiation process. The methodology of initiation estimation can be facilitated in many cases by plotting the averaged da/dN (Fig. 10b), where every point is averaged with two preceding and two following neighbouring points. The number of cycles for the initiation is estimated by summing all the cycles spent for nucleation of individual segmental cracks that occurred in the microstructural model until the observed rate drop. When combined together, the initiation and the succeeding fracture mechanics-based long crack growth allow estimation of the complete fatigue life. The combined results are typically given as the finite life region (slope region) of the *S-N* diagram.

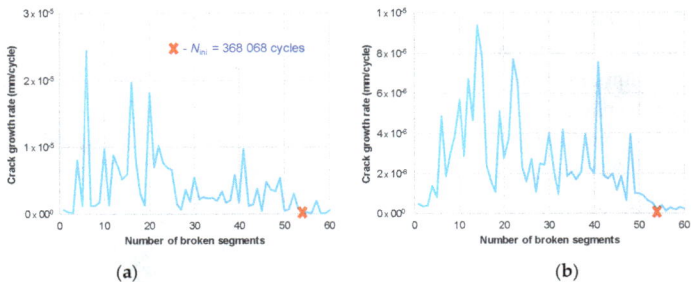

Figure 10. (a) Fatigue crack growth rate of the Fe 99.9% at the end of the crack initiation stage; (b) Averaged fatigue crack growth rate of the Fe 99.9%.

With the decrease of the loading stress amplitude, the finite life region comprises more and more of just the crack initiation stage and eventually transforms into an infinite life region, as reported by Mughrabi [62]. The aforementioned knee in the diagram, which also represents the fatigue endurance limit, can be typically recognized as the transition point between definite life (slope in the typical *S-N* curve) and infinite life (below the fatigue limit). Thus, the fatigue limit can be determined by solely calculating the initiation number of cycles as the transition of a short crack into the long crack does not take place in the fatigue limit region [62]. Accordingly, the loading amplitude can be decreased incrementally in the microstructurally-based crack initiation modelling approach, until the point where just few or no cracks appear inside the microstructural model and where extreme cycles are reached for those few nucleated cracks. An example is given for 1050A in Fig. 11, where extreme cycles as well as relatively low crack growth rates have been reached for only two nucleated cracks in the microstructural model. Such situations of the short crack retardation are considered as run-outs in the simulation analysis. Fig. 7, comprises results for fatigue life curves of all considered materials in this study. Points in Fig. 7 are average values of results from two different microstructures analyzed per material.

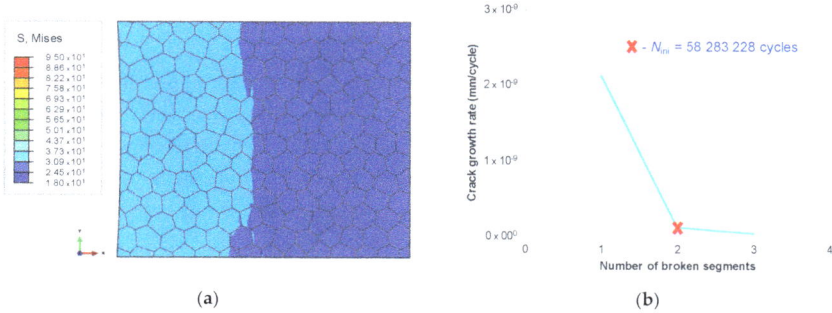

Figure 11. (a) Damaged microstructural model of the 1050A; (b) Fatigue crack growth rate of the 1050A at the end of the crack initiation stage, under 15 MPa loading stress amplitude.

The magnitudes of the fatigue (endurance) limits (S_e) of the investigated materials have been extracted from the numerically obtained S-N diagram, tabulated, and compared with the experimental values in Table 4, and further discussed in Section 4.

Table 4. Fatigue limits of the investigated materials.

Material	S_e (MPa)/Sim.	S_e (MPa)/Exp.
AISI 304	208	190–222 [63–65]
AISI 1141	152	155 [60]
Fe 99.9%	44	-
1050A	18	10-14 [66]

4. Discussion and Conclusions

As reported by several researchers, the CRSS may be up to 100 times as large in BCC steels as in metals with FCC crystal structures. However, after a detailed survey, it has been observed that there are certain FCC steels (e.g. austenitic stainless steel AISI 304) which have an unusually high CRSS. Besides of that, BCC metals typically show the "knee" and, on the other hand, certain FCC metals with a low CRSS show no sharp "knee" in the S-N diagram. Fatigue life curves in Fig. 7 illustrate the existence of definite fatigue limits in both the BCC (AISI 1141) and the FCC steels (AISI 304). According to the numerical observations, the magnitude of the CRSS is directly responsible for the occurrence or absence of the definite fatigue limit. From the microstructurally- and TM equation-based modelling point of view and from the resulting fatigue life curves in Fig. 7, the following explanations for this recognition are listed:

- The transition from infinite life (below the fatigue limit) to definite life (slope in the typical *S-N* curve) happens when a sufficient number of micro-cracks in the material microstructure (typically >10) have reached a condition for cracking. The cracking condition according to the TM equation says that the absolute magnitude of average shear stress on a grain slip band has to be higher than two times the CRSS.

- The higher the CRSS magnitude of the metal of interest, the higher the loading stress amplitude needed to accomplish the transition from the infinite to definite life, as shown in Fig. 7. In other words, the higher the CRSS magnitude, the higher the accompanying fatigue endurance limit. Despite having different crystal structures, FCC austenitic stainless steel AISI 304 and BBC vanadium-based micro-alloyed forging steel AISI 1141 have relatively high CRSS values (160 MPa and 117 MPa, respectively) and as a result fatigue limits at considerably high positions in the *S-N* diagram, Fig. 7.

- In the case of very high CRSS values, the stresses within the microstructural model are at a relatively high level at the moment of transition from infinite life to definite life, resulting in a high number of grains that are favorable for cracking. The higher the number of cracking favorable grains after the transition from infinite life to definite life, the steeper the slope of the finite life region in the *S-N* diagram.

- The study showed that the crystallographic structure is not the predominant factor that determines the shape and position of a fatigue life curve in the *S-N* diagram, but it is rather the CRSS magnitude. Namely, the higher the CRSS of a certain material, the higher the curve position is in the diagram, and the more pronounced the transition between the definite and the indefinite life region. Despite being an FCC material, the austenitic stainless steel AISI 304 showed an untypically high fatigue limit (208 MPa), which is higher than the fatigue limit of the BCC vanadium-based micro-alloyed forging steel AISI 1141 (152 MPa). The remaining two investigated FCC metals, the pure iron (Fe 99.9%) and the high purity aluminum (Al 99.5%) possess, according to this numerical study, relatively low fatigue limits, i.e. 44 and 19 MPa, respectively.

- The numerical study provided good agreement of the fatigue limits of the investigated materials with the experimentally determined fatigue limits. This observation refers firstly to AISI 1141 steel whose numerical fatigue limit of 152 MPa [52] is almost perfectly matching the experimental one, which is 155 MPa [60]. The value determined numerically for AISI 304 is fitting the span of experimental values that can be found in literature, too [63–65]. A relatively good agreement, however with a slight overestimation, has been achieved for aluminum 1050A (18 MPa vs. 10-14 MPa [66]). Reliable experimental data for Fe 99.9% could not be found. These results are summarized in Table 4 and depicted in Fig. 7.

To conclude, the analysis yields a fundamental understanding of the difference between the shapes of the fatigue life curves for steels and other metals with different crystal structures and the importance of the material parameter CRSS. The effects of crystal structure and the CRSS provide a facet of fatigue theory that is numerically predictive and which allows us to select those types of materials, which are more likely to be fatigue resistant.

References

[1]   Budynas R.G, Nisbett J.K, 2015, Fatigue failure resulting from variable loading. In Shigley's Mechanical Engineering Design, 10th ed, McGraw-Hill Education: New York, NY, USA,; pp. 273–349, ISBN 978-0-07-339820-4.

[2]   Ferro A, Montalenti G,1964, On the effect of the crystalline structure on the form of fatigue curves. Philos. Mag., 10, 1043. https://doi.org/10.1080/14786436508218923

[3]   Ferro A, Mazzetti P, Montalenti G, 1965, On the effect of the crystalline structure on fatigue: Comparison between body-centred metals (Ta, Nb, Mo and W) and face-centred and hexagonal metals. Phil. Mag. J. Theor. Exp. Appl. Phys., 12, 867–875. https://doi.org/10.1080/14786436508218923

[4]   Buck A, 1967, Fatigue properties of pure metals, Int. J. Fract. Mech. 3, 145–152. https://doi.org/10.1007/BF00182692

[5]   Grosskreutz J.C, 1971, Fatigue mechanisms in the sub-creep range. ASTM ,495, 5–60. https://doi.org/10.1520/STP26684S

[6]   Rogne B, Thaulow C, 2015, Strengthening mechanisms of iron micropillars. Phil. Mag., 95, 1814–1828. https://doi.org/10.1080/14786435.2014.984004

[7]   Uchic M.D, Dimiduk D.M, Florando J.N, Nix W.D, 2004, Sample dimensions Influence strength and crystal plasticity. Science, 305, 986–989. https://doi.org/10.1126/science.1098993

[8]   Kunz A, Pathak S, Greer J.R, 2011, Size effects in Al nanopillars: Single crystalline vs. bicrystalline. Acta Mater., 59, 4416–4424. https://doi.org/10.1016/j.actamat.2011.03.065

[9]   Greer J.R, Hosson J.T.D, 2011, Plasticity in small-sized metallic systems: Intrinsic versus extrinsic size effect. Prog. Mater Sci., 56, 654–724. https://doi.org/10.1016/j.pmatsci.2011.01.005

[10] Chen Z.M, Okamoto N.L, Demura M, Inui H, 2016, Micropillar compression deformation of single crystals of $Co_3(Al,W)$ with the $L1_2$ structure. Scr. Mater., 121, 28–31. https://doi.org/10.1016/j.scriptamat.2016.04.029

[11] Monnet G, Pouchon M.A, 2013, Determination of the critical resolved shear stress and the friction stress in austenitic stainless steels by compression of pillars extracted from single grains. Mater. Lett. 98, 128–130. https://doi.org/10.1016/j.matlet.2013.01.118

[12] Dimiduk D.M, Woodward C, LeSar R, Uchic M.D, 2006, Scale-free intermittent flow in crystal plasticity, Science 312, 1188–1190. https://doi.org/10.1126/science.1123889

[13] Csikor F.F, Motz C, Weygand D, Zaiser M, Zapperi S, 2007, Dislocation Avalanches, Strain Bursts, and the Problem of Plastic Forming at the Micrometer Scale. Science, 318, 251–254. https://doi.org/10.1126/science.1143719

[14] Okamoto N.L, Kashioka D, Inomoto M, Inui H, Takebayashi H, Yamaguchi S, 2013, Compression deformability of gamma- and zeta-Fe-Zn intermetallics to mitigate detachment of brittle intermetallic coating of galvannealed steels. Scr. Mater. 69, 307–310. https://doi.org/10.1016/j.scriptamat.2013.05.003

[15] Okamoto N.L, Fujimoto S, Kambara Y, Kawamura M, Chen Z.M.T, Matsunoshita H, Tanaka K, Inui H, George E.P, 2016, Size effect, critical resolved shear stress, stacking fault energy, and solid solution strengthening in the CrMnFeCoNi high-entropy alloy. Sci. Rep. 6, 35863. https://doi.org/10.1038/srep35863

[16] Kiener D, Motz C, Schöberl T, Jenko M, Dehm G, 2006, Determination of mechanical properties of copper at the micron scale. Adv. Eng. Mater. 8, 1119–1125. https://doi.org/10.1002/adem.200600129

[17] Jennings A.T, Burek M.J, Greer J.R, 2010, Microstructure versus Size: Mechanical properties of electroplated single crystalline Cu nanopillars. Phys. Rev. Lett. 104, 135503. https://doi.org/10.1103/PhysRevLett.104.135503

[18] Schneider A.S, Kaufmann D, Clark B.G, Frick C.P, Gruber P.A, Munig R, Kraft O, Arzt E, 2009, Correlation between critical temperature and strength of small-

scale bcc pillars. Phys. Rev. Lett. 103, 105501.
https://doi.org/10.1103/PhysRevLett.103.105501

[19]   Dimiduk D, Uchic M, Parthasarathy T, 2005, Size-affected single-slip behavior of
        pure nickel microcrystals. Acta Mater. 53, 4065–4077.
        https://doi.org/10.1016/j.actamat.2005.05.023

[20]   Greer J.R, Oliver W.C, Nix W.D, 2004, Size dependence of mechanical properties
        of gold at the micron scale in the absence of strain gradients. Acta Mater. 2005,
        53, 1821–1830. https://doi.org/10.1016/j.actamat.2004.12.031

[21]   Shade P, Uchic M, Dimiduk D, Viswanathan G, Wheeler R, Fraser H, 2011, Size-
        affected single-slip behavior of Rene N5 microcrystals. Mater. Sci. Eng. A 2012,
        535, 53–61. https://doi.org/10.1016/j.msea.2011.12.041

[22]   Okamoto N.L, Kashioka D, Hirato T, Inui H, 2014, Specimen- and grain-size
        dependence of compression deformation behavior in nanocrystalline copper. Int. J.
        Plast. 56, 173–183. https://doi.org/10.1016/j.ijplas.2013.12.003

[23]   Okamoto N.L, Inomoto M, Adachi, H, Takebayashi, H, Inui H, 2014, Micropillar
        compression deformation of single crystals of the intermetallic compound zeta-
        FeZn13. Acta Mater. 65, 229–239. https://doi.org/10.1016/j.actamat.2013.10.065

[24]   Zheng H, Cao A, Weinberger C.R, Huang J.Y, Du K, Wang J, Ma Y, Xia Y, Mao
        S.X, 2010, Discrete plasticity in sub-10-nm-sized gold crystals. Nat. Commun. 1,
        144. https://doi.org/10.1038/ncomms1149

[25]   Greer J.R, Weinberger C.R, Cai W, 2008, Comparing the strength of f.c.c. and
        b.c.c. sub-micrometer pillars: Compression experiments and dislocation dynamics
        simulations. Mater. Sci. Eng. A 493, 21–25.
        https://doi.org/10.1016/j.msea.2007.08.093

[26]   Schneider A, Clark B, Frick C, Gruber P, Arzt E, 2009, Effect of orientation and
        loading rate on compression behavior of small-scale Mo pillars. Mater. Sci. Eng.
        A 508, 241–246. https://doi.org/10.1016/j.msea.2009.01.011

[27]   Hagen A, Thaulow C, 2016, Low temperature in-situ micro-compression testing of
        iron pillars, Mater. Sci. Eng. A 678, 355–364.
        https://doi.org/10.1016/j.msea.2016.09.110

[28]   Bruesewitz C, Knorr I, Hofsaess H, Barsoum M.W, Volkert C.A, 2013, Single
        crystal pillar microcompression tests of the MAX phases Ti_2InC and Ti_4AlN_3, Scr.
        Mater. 69, 303–306. https://doi.org/10.1016/j.scriptamat.2013.05.002

[29] Feller-Kniepmeier M, Hundt M, 1983, Deformation properties of high purity alpha-Fe single crystals. Scr. Metall. 17, 905–908. https://doi.org/10.1016/0036-9748(83)90259-4

[30] Stein D, Low J, Seybolt A, 1963, The mechanical properties of iron single crystals containing less than 5×10^{-3} ppm carbon. Acta Metall. 11, 1253–1262. https://doi.org/10.1016/0001-6160(63)90114-7

[31] Stein D.F, Low J.R, 1966, Effects of orientation and carbon on the mechanical properties of iron single crystals. Acta Metall. 14, 1183–1194. https://doi.org/10.1016/0001-6160(66)90236-7

[32] Guo E.-Y, Xie H X, Singh S. S, Kirubanandham A, Jing T, Chawla N, 2014, Mechanical characterization of microconstituents in a cast duplex stainless steel by micropillar compression. Mater. Sci. Eng. A 598, 98–105. https://doi.org/10.1016/j.msea.2014.01.002

[33] Cruzado A, Gan B, Jimenez M, Barba D, Ostolaza K, Linaza A, Molina-Aldareguia J, Llorca J, Segurado J, 2015, Multiscale modeling of the mechanical behavior of IN718 superalloy based on micropillar compression and computational homogenization. Acta Mater. 98, 242–253. https://doi.org/10.1016/j.actamat.2015.07.006

[34] Jin H.H, Ko E, Kwon J, Hwang S.S, Shin C, 2016, Evaluation of critical resolved shear strength and deformation mode in proton-irradiated austenitic stainless steel using micro-compression tests. J. Nucl. Mater. 470, 155–163. https://doi.org/10.1016/j.jnucmat.2015.12.029

[35] Wu J, Tsai W, Huang J, Hsieh C, Huang G.-R, 2016, Sample size and orientation effects of single crystal aluminum. Mater. Sci. Eng. A 662, 296–302. https://doi.org/10.1016/j.msea.2016.03.076

[36] Palomares Garcia A.J, Perez-Prado M.T, Molina-Aldareguia J.M, 2017, Effect of lamellar orientation on the strength and operating deformation mechanisms of fully lamellar TiAl alloys determined by micropillar compression. Acta Mater. 123, 102–114. https://doi.org/10.1016/j.actamat.2016.10.034

[37] Campos M, Bautista A, Caceres D, Abenojar J, Torralba J, 2003, Study of the interfaces between austenite and ferrite grains in P/M duplex stainless steels. J. Eur. Ceram. Soc. 23, 2813–2819. https://doi.org/10.1016/S0955-2219(03)00293-0

[38] Ramazani A, Mukherjee K, Prahl U, Bleck W, 2012, Modelling the effect of microstructural banding on the flow curve behaviour of dual-phase (DP) steels. Comput. Mater. Sci. 52, 46–54. https://doi.org/10.1016/j.commatsci.2011.05.041

[39] Hocker S, Schmauder S, Bakulin A.V, Kulkova S.E, 2014, Ab initio investigation of tensile strengths of metal(1 1 1)/alpha-Al_2O_3(0 0 0 1) interfaces. Philos. Mag. 94, 265–284. https://doi.org/10.1080/14786435.2013.852288

[40] Kulkova S.E, Bakulin A.V, Kulkov S.S, Hocker S, Schmauder S, 2015, Influence of interstitial impurities on the Griffith work in Ti-based alloys. Phys. Scr. 90, 094010. https://doi.org/10.1088/0031-8949/90/9/094010

[41] Bakulin A.V, Spiridonova T, Kulkova S.E, Hocker S, Schmauder, S, 2016, Hydrogen diffusion in doped and undoped alpha-Ti: An ab-initio investigation. Int. J. Hydrogen Energy 41, 9108–9116. https://doi.org/10.1016/j.ijhydene.2016.03.192

[42] Kohler C, Kizler P, Schmauder S, 2005, Atomistic simulation of precipitation hardening in alpha-iron: Influence of precipitate shape and chemical composition. Modell. Simul. Mater. Sci. Eng. 13, 35–45. https://doi.org/10.1088/0965-0393/13/1/003

[43] Prskalo A.P, Schmauder S, Ziebert C, Ye J, Ulrich S, 2010, Molecular dynamics simulations of the sputtering of SiC and Si_3N_4. Surf. Coat. Technol. 204, 2081–2084. https://doi.org/10.1016/j.surfcoat.2009.09.043

[44] Prskalo A.-P, Schmauder S, Ziebert C, Ye J, Ulrich S, 2011, Molecular dynamics simulations of the sputtering process of silicon and the homoepitaxial growth of a Si coating on silicon. Comput. Mater. Sci. 50, 1320–1325. https://doi.org/10.1016/j.commatsci.2010.08.006

[45] Bozic Z, Schmauder S, Mlikota M, Hummel M, 2014, Multiscale fatigue crack growth modelling for welded stiffened panels. Fatigue Fract. Eng. Mater. Struct., 37, 1043–1054. https://doi.org/10.1111/ffe.12189

[46] Bozic Z, Schmauder S, Mlikota M, Hummel M, 2018, Multiscale fatigue crack growth modeling for welded stiffened panels. In Handbook of Mechanics of Materials, Schmauder S, Chen C.-S, Chawla, K.K, Chawla N, Chen W, Kagawa Y, Eds, Springer: Singapore, pp. 1–21, ISBN 978-981-10-6855-3. https://doi.org/10.1007/978-981-10-6855-3_73-1

[47] Tanaka K, Mura T,1981, A dislocation model for fatigue crack initiation. J. Appl. Mech. 48, 97–103. https://doi.org/10.1115/1.3157599

[48] Tanaka K, Mura T, 1982, A theory of fatigue crack initiation at inclusions. Metall. Trans. A ,13, 117–123. https://doi.org/10.1007/BF02642422

[49] Mlikota M, Staib S, Schmauder S, Bozic Z. 2017, Numerical determination of Paris law constants for carbon steel using a two-scale model. J. Phys. Conf. Ser. 843, 012042. https://doi.org/10.1088/1742-6596/843/1/012042

[50] Mlikota M, Schmauder S, Bozic Z, Hummel M, 2017, Modelling of overload effects on fatigue crack initiation in case of carbon steel. Fatigue Fract. Eng. Mater. Struct. 40, 1182–1190. https://doi.org/10.1111/ffe.12598

[51] Mlikota M, Schmauder S, 2017, Numerical determination of component Woehler curve. DVM Bericht 1684, 111–124

[52] Mlikota M, Schmauder S, Bozic Z, 2018, Calculation of the Woehler (S-N) curve using a two-scale model. Int. J. Fatigue 114, 289–297. https://doi.org/10.1016/j.ijfatigue.2018.03.018

[53] ABAQUS, version 2018, Abaqus Documentation, Simulia: Providence, RI, USA,.

[54] Socie D.F. 1996, Fatigue damage simulation models for multiaxial loading. In Proceedings of the Sixth International Fatigue Congress (Fatigue '96), Berlin, Germany, 6–10 May; pp. 967–976. https://doi.org/10.1016/B978-008042268-8/50038-1

[55] Glodez S, Jezernik N, Kramberger J, Lassen T, 2010, Numerical modelling of fatigue crack initiation of martensitic steel. Adv. Eng. Software 2010, 41, 823–829. https://doi.org/10.1016/j.advengsoft.2010.01.002

[56] Jezernik N, Kramberger J, Lassen T, Glodez S, 2010, Numerical modelling of fatigue crack initiation and growth of martensitic steels. Fatigue Fract. Eng. Mater. Struct. 33, 714–723. https://doi.org/10.1111/j.1460-2695.2010.01482.x

[57] Huang X, Brueckner-Foit A, Besel M, Motoyashiki Y, 2007, Simplified three-dimensional model for fatigue crack initiation. Eng. Fract. Mech. 74, 2981–2991. https://doi.org/10.1016/j.engfracmech.2006.05.027

[58] Briffod F, Shiraiwa T, Enoki M, 2016, Fatigue crack initiation simulation in pure iron polycrystalline aggregate. Mater. Trans. 57, 1741–1746. https://doi.org/10.2320/matertrans.M2016216

[59] Paris P, Erdogan F, 1963, A critical analysis of crack propagation laws. J. Basic Eng. 85, 528–533. https://doi.org/10.1115/1.3656900

Materials Research Forum LLC
https://doi.org/10.21741/9781644901656-3

[60] Fatemi A, Zeng Z, Plaseied A, 2003,Fatigue behavior and life predictions of notched specimens made of QT and forged microalloyed steels, Int. J. Fatigue 2004, 26, 663–672. https://doi.org/10.1016/j.ijfatigue.2003.10.005

[61] Newman J, Phillips E, Swain M, 1999, Fatigue-life prediction methodology using small-crack theory. Int. J. Fatigue. 21, 109–119. https://doi.org/10.1016/S0142-1123(98)00058-9

[62] Mughrabi H, 2015, Microstructural mechanisms of cyclic deformation, fatigue crack initiation and early crack growth. Philos. Trans. R. Soc. Lond. Ser. A 373. https://doi.org/10.1098/rsta.2014.0132

[63] Boyer H, 1985, Atlas of Fatigue Curves, 1st ed, American Society for Metals: Materials Park, OH, USA,; ISBN 0871702142.

[64] Atzori B, Meneghetti G, Ricotta M, 2011, Analysis of the fatigue strength under two load levels of a stainless steel based on energy dissipation. Fract. Struct. Integrity 17, 15–22. https://doi.org/10.3221/IGF-ESIS.17.02

[65] Ben Fredj, N Ben Nasr M, Ben Rhouma A, Sidhom H, Braham C, 2004, Fatigue life improvements of the AISI 304 stainless steel ground surfaces by wire brushing. J. Mater. Eng. Perform. 13, 564–574. https://doi.org/10.1361/15477020420819

[66] Sakin R, 2018, Investigation of bending fatigue-life of aluminum sheets based on rolling direction. Alex. Eng. J. 57, 35–47. https://doi.org/10.1016/j.aej.2016.11.005

Materials Research Forum LLC
https://doi.org/10.21741/9781644901656-4

Chapter 4

A Newly Discovered Relation between the Critical Resolved Shear Stress and the Fatigue Endurance Limit for Metallic Materials

M. Mlikota[1], S. Schmauder[1]

[1]Institute for Materials Testing, Materials Science and Strength of Materials (IMWF), University of Stuttgart, Pfaffenwaldring 32, 70569 Stuttgart, Germany

Abstract

The chapter introduces a valuable new description of fatigue strength in relation to material properties and thus a new perspective on the overall understanding of the fatigue process. Namely, a relation between the endurance limits and the accompanying values of the critical resolved shear stress (CRSS) for various metallic materials has been discovered by means of a multiscale approach for fatigue simulation. Based on the uniqueness of the relation, there is a strong indication that it is feasible to relate the endurance limit to the CRSS and not to the ultimate strength, as often done in the past.

Keywords

Multiscale Simulation, Fatigue, Metals, CRSS, Endurance Limit

Contents

Originally published in the Journal of Metals (2020), 10, 803
https://doi.org/10.3390/met10060803

1.   Introduction

A scientific approach to the question of fatigue strength would be to consider the effects of crystal structure on fatigue mechanisms [1–4]. Researchers from the field of fatigue are aware of the ratio between endurance limit and ultimate tensile strength, S_e/R_m (see Fig. 1b). This ratio is also known as the fatigue ratio and is typically higher for ferrous materials (including steels—see red line in Fig. 1b), which are of the body-centered cubic (BCC) type, than for non-ferrous materials (see, e.g., blue line in Fig. 1b), which possess a face-centered cubic (FCC) crystallographic structure. Furthermore, ferrous materials generally show a pronounced "knee" in the strength-life (S-N, or Wöhler) diagram at about 10^6 cycles, after which the fatigue life curve flattens (Fig. 2). The fatigue strength at this point is known as the endurance limit (S_e). Interestingly, non-ferrous materials exhibit a gradual flattening between 10^7–10^8 cycles. Although some researchers have explained these effects in terms of strain ageing and dislocation locking [4], there is also evidence that the involved crystal structure plays an important role [1,2].

However, the study from [5] showed that the crystallographic structure is not the predominant factor that determines the shape and position of the fatigue life curves in the S-N diagram, but it is rather the parameter critical resolved shear stress (CRSS). Mlikota and Schmauder [5] reported the existence of a pronounced transition from finite life (slope in the typical S-N curve; see Fig. 2) to the infinite life region (below the endurance limit) as well as high S_e values even in some FCC metals with relatively high CRSS magnitudes. Namely, the higher the CRSS of a certain material, the more pronounced is the transition between the finite and the infinite life region, the higher is the curve position in the diagram and accordingly the higher is its S_e magnitude [5]. The present study is the follow-up study of the one published in [5] and brings new insights to the simulation-based understanding of CRSS for the fatigue performance of metallic materials.

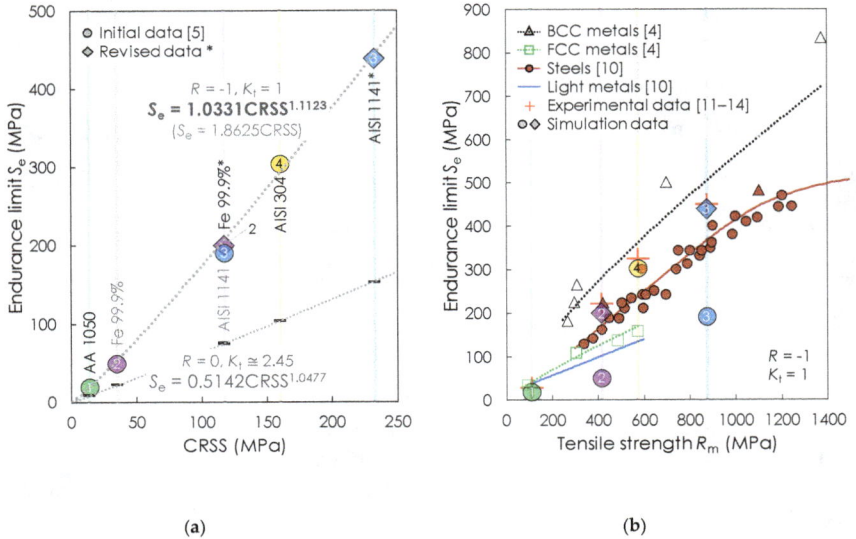

(a)

(b)

Figure 1. (a) Relation between endurance limit (S_e) values (from [5] and updated with the new results for the Fe 99.9% and for the steel AISI 1141, marked with diamonds and with *) of the investigated metals and their critical resolved shear stress (CRSS) values [6–9]. (b) Relation between S_e and ultimate tensile strength (R_m) values for various metals [4,10–14], including new simulation-based S_e values (diamonds) and those from [5] (circles). Simulation-based S_e values for the standard loading case characterized by loading ratio $R = -1$ and stress concentration factor $K_t = 1$ are obtained from the values characteristic for $R = 0$ and $K_t \cong 2.45$ by using an approach taking into account mean stress and/or loading ratio [10] in combination with the notch sensitivity approach [15] (see Appendix A for more details).

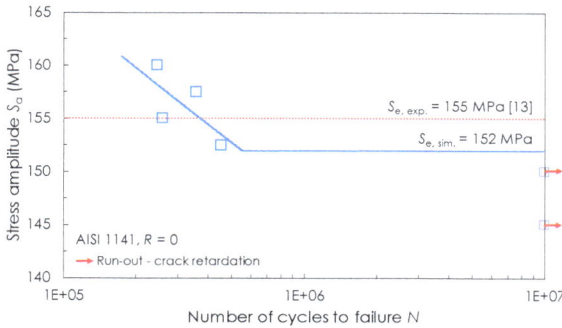

Figure 2. Comparison between simulated and experimental S_e values for AISI 1141 steel [13]. For details on the numerical determination of S_e values, see [5,16,17].

2. Methods and Materials

2.1 Methods and Reference to Previous Work

A study published in [5] by Mlikota and Schmauder dealt with the numerical estimation of the fatigue life represented in the form of *S-N* curves of metals with BCC and FCC crystallographic structures and with different magnitudes of CRSS. An example of a simulation-based *S-N* curve is shown in Fig. 2 for BCC steel AISI 1141. Such fatigue life curves are obtained by determining the number of cycles for initiation of a short crack under the influence of microstructure (N_{ini}; Fig. 3b) and subsequent number of cycles for the growth of a long crack (N_{prop}; Fig. 3c), respectively. Final failure of a specimen or a component occurs at the number of cycles $N_f = N_{ini} + N_{prop}$. Micro-models containing microstructures of the materials are set up by using the finite element method (FEM) and are analyzed in combination with the Tanaka-Mura (TM) equation [18,19] in order to estimate the number of cycles required for crack initiation (see Fig. 3b and especially [5] for more details). Long crack growth analysis is typically based on classical fracture mechanics.

A dislocation model forms the physical basis of the TM equation (Equation (1)), which is frequently used to determine when a grain, subjected to an outer cyclic loading, will develop a slip band and subsequently a micro-crack. The number of cycles, N_g, needed for micro-crack nucleation within a single grain can be derived as follows [18,19]:

Multiscale Fatigue Modelling of Metals   Materials Research Forum LLC
Materials Research Foundations **114** (2022) 66-82   https://doi.org/10.21741/9781644901656-4

$$N_s = \frac{8GW_c}{(1-v)(\Delta\tau_s - 2\text{CRSS})^2 \pi d_s} \tag{1}$$

One of the parameters of the TM model (Eq. 1) is the CRSS, which is a threshold value of the shear stress along the glide direction that a dislocation needs to surpass in order to start moving. According to the TM model, micro-cracks form along slip bands (Fig. 3b), depending on grain size (i.e., slip band length) d, the average shear stress range $\Delta\tau$ on the slip band, the shear modulus G, the crack initiation energy W_c, Poisson's ratio v, and the CRSS [16,17,20–22]. A more extended and detailed description of the implementation of the TM equation into FEM-based modelling and simulation of the crack initiation process has been reported in publications of the authors of this study in [5,16,17,20,23–25] and by other researchers, too, in [21,22,26,27].

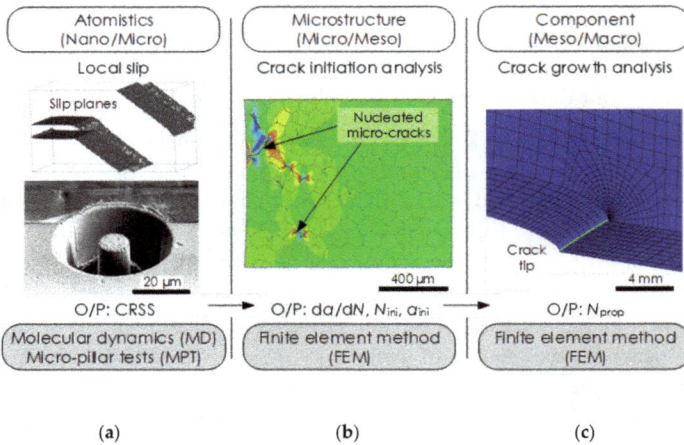

Figure 3. Multiscale approach—coupling of methodologies at the relevant scales and accompanying outputs (O/P). (a) Determination of the parameter CRSS either from molecular dynamics (MD) simulations or from micro-pillar tests (MPT). (b) Determination of crack growth rate (da/dN) and number of stress cycles for crack initiation (N_{ini}) from crack initiation analysis. (c) Determination of number of stress cycles for crack propagation (N_{prop}) [5,8,23,30].

The multiscale approach for fatigue simulation, consisting of CRSS determination either from micro-pillar tests (MPT) or from molecular dynamics (MD) simulations (both in Fig. 3a), crack initiation simulations based on the TM equation (Fig. 3b) and long crack growth simulations based on fracture mechanics principles (Fig. 3c), has been previously reported in [8,16,24,25,28,29]. The methodology forms the basis of the past [5] and present research work, which has been conducted with the aim to further elucidate the relevance of the parameter CRSS for fatigue strength in general.

2.2   Materials

Table 1, contains metallic materials considered in the study from [5] and their mechanical properties, namely, the Young's modulus E, the shear modulus G, the Poisson's ratio v, the yield strength $R_{p0.2}$, the ultimate strength R_m, their average grain size d, and eventually the CRSS values. The details on boundary and loading conditions as well as on specimen geometry that were applied in the study can be found in [5], too. It is suitable to indicate here that all the materials' constitutive laws have been defined as purely elastic, i.e., just by using the elastic material constants. In another study of Mlikota and Schmauder [17] on aluminum alloy AlSi8Cu3, it has been shown that plasticity does influence the fatigue performance of this alloy, however, not significantly.

Table 1. Mechanical properties of the considered metals.

Material	E (GPa)	G (GPa)	v	$R_{p0.2}$ (MPa)	R_m (MPa)	d (μm)	CRSS (MPA)
AISI 304	188	79.0	0.26	322 [31]	574 [31]	30	160 [9]
AISI 1141	200	78.125	0.28	564 [13]	875 [13]	60	117 [8]
Fe 99.9%	205	81.0	0.28	260 [32]	414 [34]	65	35 [7]
AA 1050	72	26.0	0.33	95 [33]	110 [33]	65	14 [6]

2.3   New Insights into Previous Work

As reported by several researchers, the CRSS may be up to 100 times larger in BCC steels than in metals with FCC crystal structures [4]. However, after a detailed survey, it was observed that certain FCC steels also exist which have an unusually high CRSS (e.g., austenitic stainless steel AISI 304 with a 160 MPa high CRSS, see Table 1). As already mentioned, BCC metals typically show a pronounced transition from finite life to the endurance limit region, and on the other hand, certain FCC metals with a low CRSS show relatively smoother transition between the two regions in the S-N diagram. In addition to that, the results from [5] illustrate (Table 2) the existence of definite endurance limits in the considered BCC (AISI 1141 and Fe 99.9%) as well as FCC steels (AISI 304 and AA

1050). Despite being an FCC material, the austenitic stainless steel AISI 304 shows an untypically high S_e value of 103 MPa (expressed in terms of nominal stress amplitude S_a for loading ratio $R = 0$ and stress concentration factor $K_t \cong 2.45$), which is higher than the S_e value estimated for the BCC steel AISI 1141 (76 MPa for CRSS = 117 MPa). The remaining two investigated metals, the BCC-based pure iron (Fe 99.9%) and the FCC-based high purity aluminum alloy (AA) 1050 possess, according to the numerical study from [5], relatively low endurance limits, i.e., 22 and 8 MPa, respectively.

According to the observations from [5], the magnitude of the CRSS seems to be directly responsible for the magnitude of the simulated S_e values of all investigated materials. The simulation-based magnitudes of S_e from [5] have been tabulated in Table 2 (see fourth column). Furthermore, these values from the fourth column of Table 2 have been converted to the case of an unnotched sample ($K_t = 1$) with purely alternating stress ($R = -1$) in order to compare them with the experimental S_e values; see fifth and sixth columns, respectively. It is opportune to indicate that the authors realized in the time after publication of the first paper [5] on the CRSS relevance for the fatigue performance of metallic materials and before performing the present study that the values of S_e in [5] represent maximum nominal stress (S_{max}) and not nominal stress amplitude (S_a) as given there. Since the R ratio selected in the present simulations was 0 (for axial loading and $K_t \cong 2.45$), the revised S_e values in Table 2 (fourth column) are two times lower than those published in [5] ($S_a = S_{max}/2$ for $R = 0$).

Table 2. Simulation-based endurance limits (S_e) from [5] of the investigated materials in comparison with experimental values (S_e values are expressed in terms of nominal stress amplitude S_a). For details on how to deduct S_e values for $R = -1$ and $K_t = 1$ (axial loading), see Appendix A.

Material	Lattice	CRSS (MPa)	S_e (MPa)/Sim. $R = 0$, $K_t \cong 2.45$	S_e (MPa)/Sim. $R = -1$, $K_t = 1$	S_e (MPa)/Exp. $R = -1$, $K_t = 1$
AISI 304	FCC	160 [9]	103 [5]	303.7	325 [14]
AISI 1141	BCC	117 [8]	76 [5]	190.7	450 [13]
Fe 99.9%	BCC	35 [7]	22 [5]	49.7	222 [12]
AA 1050	FCC	14 [6]	8 [5]	19.5	29 [11]

As visible in Table 2, the numerical study provided relatively good agreement between the calculated S_e values and the experimentally determined S_e values of some investigated materials (AISI 304 and AA 1050). This observation refers firstly to AISI 304 steel whose numerically determined S_e value of 303.7 MPa well fits the experimental value of 325 MPa that can be found in literature [14]. A relatively good agreement, however with a slight underestimation, was achieved for the aluminum alloy AA 1050 (19.5 versus 29 MPa [11]).

An exception is the value determined for the steel AISI 1141 of 190.7 MPa, which is considerably lower than the experimental counterpart (450 MPa [13]). It is expected that a better agreement with the experimental results can be achieved by selecting another— considerably higher—CRSS magnitude. The CRSS of 117 MPa calculated by means of MD for BCC α-Fe by Hummel [8] was used in the initial study [5] due to the lack of a more appropriate value for the steel AISI 1141. The reason to expect a higher CRSS magnitude— and by that a higher S_e value—for the considered steel, AISI 1141, are its improved mechanical properties over iron—achieved by microstructural modifications, i.e., by the addition of small amounts of the micro-alloying elements such as vanadium (V, 0.053 wt%) [13,35]. It is known that such alloying elements contribute to the strength in general, but also to the CRSS magnitude. Another reason to expect a higher CRSS is the relatively high $R_{p0.2}$ of around 560 MPa for this steel. Another divergence from the values that can be found in literature is seen for Fe 99.9%; the numerically obtained S_e value of 49.7 MPa is considerably lower than the experimental value of 222 MPa, as reported in [12]. Here again the CRSS magnitude could be considered as a reason for the discrepancy. Namely, the CRSS value of 35 MPa extracted from MPT of Rogne and Thaulow [7] and used in [5] is considerably lower than the MD-based 117 MPa [8] for BCC α-Fe.

Interestingly, when these four simulation-based S_e values of the investigated metals (Table 2) are plotted versus their R_m values (Table 1), again a similar observation from Figure 1b follows; namely, the S_e-R_m points of AA 1050 (green circle denoted with number 1) and AISI 304 (golden circle denoted with number 4)—agreeing well with the S_e experiments in Table 2—fall into the range of points characteristic for their groups of materials, i.e., to light metals and steels, respectively. On the other hand, the S_e-R_m points of the two other metals—Fe 99.9% (purple circle denoted with number 2 in Fig. 1b) and AISI 1141 (blue circle denoted with number 3 in Fig. 1b)—deviate considerably from the data representing steels and BCC metals, the same as they deviate from the experimental S_e values from Table 2 (see red crosses in Fig.1b). This observation suggests that the S_e values and by that the CRSS values, too, of the last two metals might be too low. It is necessary to point out once again that simulation-based S_e values for the standard loading case characterized by loading ratio $R = -1$ and stress concentration factor $K_t = 1$ (Fig.1b) are obtained from the S_e values characteristic for $R = 0$ and $K_t \cong 2.45$ (direct results from the simulations) by using an approach taking into account mean stress and/or loading ratio [10] in combination with the notch sensitivity approach [15] (see Figure 1a as well as Appendix A for more details).

3. Results

3.1 Relation between the Critical Resolved Shear Stress and the Fatigue Endurance Limit

When the numerically determined S_e values of the four investigated metals from the initial study published in [5] are plotted with respect to their initially prescribed CRSS values (see both values in Table 2), an interesting relation can be observed, as shown in Fig. 1a. Namely, the S_e and CRSS values relate to each other in a linear manner, despite considering incorrect CRSS values in some cases (Fe 99.9% and AISI 1141). This newly discovered relation can be expressed by a power-law:

$$S_e = m_0 CRSS^s \tag{2}$$

where m_0 is the intercept with the y-axis in Fig. 1a and s the slope of the dotted lines. According to the power-law approximation, m_0 equals 0.5142 and s is 1.0477 for loading ratio $R = 0$ and stress concentration factor $K_t \cong 2.45$. Interestingly, the slope factor s is approximately equal to 1 in this case. The S_e-CRSS relation can also be expressed by using a linear function, as $S_e = 0.65$CRSS for these specific R (= 0) and K_t ($\cong 2.45$) values. However, the power-law approximation is preferably used due to higher accuracy, while the simple linear relationship is practically simpler and easier to use. It is noteworthy that the parameters m_0 and s of Eq. 2 are dependent on R ratio, K_t, etc. For the standard loading case ($R = -1$ and $K_t = 1$), $m_0 = 1.0331$ and $s = 1.1123$, while the slope of the linear function is equal to 1.8625 (Fig. 1a).

3.2 Application of the Newly Discovered Relation

Even though based purely on the simulation results and just partly validated, the S_e-CRSS relation (Eq. 2) can be used as a valuable tool in the next step to shed some light on the two cases in Table 2 (Fe 99.9% and AISI 1141) where the discrepancies with respect to the experimental results have been observed. A rather straightforward case to clarify is the one of Fe 99.9%, where by using the CRSS = 117 MPa [8] for BCC α-Fe directly in the S_e-CRSS relation (for $R = 0$ and $K_t \cong 2.45$), an endurance limit of 75.5 MPa is obtained. To prove the approach also from the numerical side, an additional simulation with the CRSS value of 117 MPa was performed, resulting in an S_e value of 74 MPa, which matches well the experimental $S_e = 222$ MPa [12] when translated to the case characterized by $R = -1$ and $K_t = 1$. This could be a confirmation that the initially used [5] and MPT-based CRSS of just 35 MPa [7] is too low. The other case of AISI 1141 can be approached from another side; namely, by knowing the target S_e = 155 MPa from the experimental study of Fatemi et al. [13] (R = 0 and $K_t \cong 2.45$), an estimation of the necessary CRSS to reach this S_e value using the multiscale fatigue simulation approach (Fig. 3) can be done by means of the S_e-CRSS relation, CRSS= 232.5 MPa. To validate this estimation, is was necessary to perform additional simulations to determine the endurance limit by taking all the input parameters the same as in the study on AISI 1141 from [5] and just by replacing the previously used CRSS of 117 MPa [8] with the new S_e-CRSS relation-based value of 232.5 MPa. The failure cycles resulting from the S_e simulations are presented in Fig. 2 versus the applied amplitude levels (S_a = 145–160 MPa, R = 0) and are, at the same place, compared with the experimental result. These results confirm the estimation of the CRSS magnitude of 232.5 MPa for the steel AISI 1141 by using the S_e-CRSS relation as being correct; the numerically obtained S_e = 152 MPa is slightly lower than the experimentally derived S_e = 155 MPa [13] (= 450 MPa for R = −1 and K_t = 1; Table 3), which seems to be an acceptable deviation of 2% only.

Moreover, data from Table 2 can now be revised with the new results for the steel AISI 1141 and for Fe 99.9% in Table 3. Aside from that, these new simulation-based results are added to Figure. 1 and at the same place visually compared with the experimental values for these metals (see red crosses in Fig. 1b).

*Table 3. Simulation-based S_e values of the investigated materials (revised with the new results for the steel AISI 1141 and for the Fe 99.9%, marked with * in comparison with experimental values. For details on how to deduct S_e values for R = -1 and K_t = 1, see Appendix A).*

Material	R_m (MPa)	CRSS (MPa)	S_e (MPa)/Sim. R = 0, $K_t \cong 2.45$	S_e (MPa)/Sim. R = −1, K_t = 1	S_e (MPa)/Exp. R = −1, K_t = 1
AISI 304	574 [31]	160 [9]	103 [5]	303.7	325 [14]
AISI 1141*	875 [13]	232.5 (Eqn.2)	152	438.5	450 [13]
AISI 1141	875 [13]	117 [8]	76 [5]	190.7	450 [13]
Fe 99.9%*	414 [34]	117 [8]	74	200	222 [12]
Fe 99.9%	414 [34]	35 [7]	22 [5]	49.7	222 [12]
AA 1050	110 [33]	14 [6]	8 [5]	19.5	29 [11]

It is important to note that no matter whether there is a correct (i.e., validated by a correct resulting S_e value of an investigated existing material) or incorrect CRSS value, there is always a linear relation between the used CRSS value and the numerically obtained S_e value, as can be seen in Fig. 1a.

In addition, and in contrast to the initial data from Fig. 1b (see circles denoted with numbers 2 and 3), the S_e-R_m points of the two revised metals—Fe 99.9% (purple diamond denoted with number 2) and AISI 1141 (blue diamond denoted with number 3)—agree well with the experimental values for these metals (see red crosses) as well as falling into the acceptable range of points characteristic for steels and BCC metals. This suggests that the revised S_e values of these two metals should be correct—as well as the accompanying CRSS values—similar to those of AA 1050 (green circle denoted with number 1 in Fig.1b) and AISI 304 (golden circle denoted with number 4 in Fig. 1b), which were already considered correct in the initial study from [5].

4.   Discussions

It follows from these observations that the S_e-CRSS relation (Eq. 2) introduces a valuable new description of fatigue strength relations in material properties and a new perspective on the overall understanding of the fatigue process, especially in comparison to contemporary relations where S_e is being related to, e.g., R_m in a non-unique manner (Fig. 1b). Accordingly, it seems to be more logical to relate S_e to CRSS and not to R_m due to the scattering of R_m that results from different strain hardening levels (i.e., cold-working) of the material [36], in addition to other strengthening mechanisms like grain boundary strengthening and phase boundary strengthening, which on the other hand have no influence on the CRSS. (see [5] for more details on the known and applicable linear super positioning principle of the strengthening mechanisms that contribute to the CRSS magnitude).

5.   Conclusions

To conclude, the presented analysis yields a groundbreaking view on the importance of the parameter critical resolved shear stress (CRSS) for estimating the fatigue strength of metallic materials. The newly discovered linear relation between the endurance limit (S_e) and CRSS provides a facet of fatigue theory which is numerically predictive and which allows the selection of fatigue resistant materials. Even though additional simulations as well as experimental studies are planned to uphold this finding, the S_e-CRSS relation can already now be used to estimate endurance limits of metallic materials solely from their CRSS values—which can be on the other hand estimated from micro-pillar tests, from molecular dynamics simulations or by using the linear super positioning principle of the strengthening mechanisms that contribute to its magnitude.

Appendix A

The endurance limit (S_e) value for the stress concentration factor (K_t) equal to 1 (i.e., notch radius $r = 0$ mm—unnotched sample) can be deduced from any S_e value determined at $K_t > 1$ (i.e., $r > 0$) by multiplying it by a factor K_{fat}, which is commonly called fatigue stress concentration factor, i.e., [15,37]:

$$S_{e,Kt1} = S_e K_{fat} \tag{A1}$$

The factor K_{fat} is determined from the factor K_t by using the expression

$$K_{fat} = 1 + q(K_t - 1) \tag{A2}$$

where q is the notch sensitivity and can be obtained for different types of metals from a diagram q versus r, as shown in Figure A1 [15].

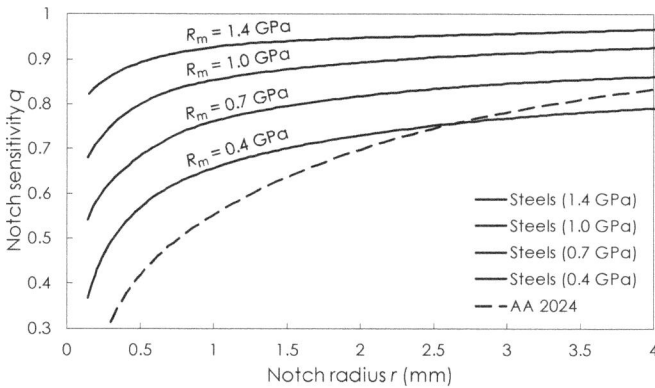

Figure A1. Notch sensitivity charts for steels with different ultimate strengths (R_m) and aluminum alloys AA 2024 subjected to reversed bending or reversed axial loads. For larger notch radii, the use of the values of q corresponding to r = 4 mm is recommended [15].

Furthermore, any S_e value determined for a non-zero mean stress ($S_m \neq 0$, i.e., $R \neq -1$) can be translated to the zero S_m case ($S_m = 0$, $R = -1$; often referred to as the standard loading case) as follows [10]:

$$S_{e,R-1} = S_e \sqrt{[1 - (S_m/R_m)]} \tag{A3}$$

By combining these two approaches (Equations (A1) and (A3)), it can explained how the S_e values characterizing purely alternating stress ($R = -1$) and for an unnotched sample ($K_t = 1$) from Tables 2 and 3 were determined. Table A1 shows the translation of the simulation-based S_e values of the investigated metals from loading and geometry conditions defined by $R = 0$ and $K_t \cong 2.45$ values to the $R = -1$ and $K_t = 1$ case.

Table A1. Simulation-based S_e values of the investigated materials (revised with the new results for the steel AISI 1141 and for the Fe 99.9%, marked with *, see Section 3 for more details) translated to the case of an unnotched sample ($K_t = 1$) and of purely alternating stress ($R = -1$).

Material	S_e (MPa) $R = 0$ $K_t \cong 2.45$	K_t	q	K_{fat}	S_e (MPa) $R = 0$ $K_t = 1$	S_m (MPa)	R_m (MPa)	S_e (MPa) $R = -1$ $K_t = 1$ (Axial)
AISI 304	103 [5]	2.53	0.83	2.27	233.8	233.8	574 [31]	303.7
AISI 1141 *	152	2.39	0.90	2.25	342.2	342.2	875 [13]	438.5
AISI 1141	76 [5]	2.39	0.90	2.25	171.1	171.1	875 [13]	190.7
Fe 99.9% *	74	2.41	0.80	2.13	157.5	157.5	414 [34]	200.0
Fe 99.9%	22 [5]	2.41	0.80	2.13	46.8	46.8	414 [34]	49.7
AA 1050	8 [5]	2.46	0.84	2.23	17.8	17.8	110 [33]	19.5

[A1] *Stress concentration factors (K_t) are determined from the numerical model of the notched sheet sample from [13] and vary between each material slightly due to different material properties defining their stress-strain responses. $K_t = 2.45$ is the average value.*

The translation of the experimental S_e values (from Tables 2 and 3) of the investigated metals from loading and geometry conditions defined by different R and K_t values to the $R = -1$ and $K_t = 1$ case is shown in Table A2.

Table A2. Experimental S_e values of the investigated materials translated to the case defined by $K_t = 1$ and $R = -1$.

Material	S_e (MPa)	R	L	K_t	q	K_{fat}	S_e (MPa) $R = 0$ $K_t = 1$	S_m (MPa)	R_m (MPa)	S_e (MPa) $R = -1$ $K_t = 1$ (Axial)
AISI 304	217 [14]	−1	a	1.6	0.83	1.50	->	-	574 [31]	325
AISI 1141	155 [13]	0	a	2.39	0.90	2.25	348.9	348.9	875 [13]	450
Fe 99.9%	150 [12]	−1	a	1.6	0.80	1.48	->	-	414 [34]	222
AA 1050	34.5 [11]	−1	b	1	-	-	->	-	110 [33]	29

[A2] *Stress concentration factor of the AISI 1141 steel ($K_t = 2.39$) is determined from the numerical model of the notched sheet sample from [13] and varies slightly from the value reported in the same source ($K_t = 2.75$). The K_t values for the notched sheet specimens of AISI 304 steel (see [14]) and of Fe 99.9% (see [12]) are determined as recommended in [15]—see page 1034, Figure A-15-3. L—loading type; a—axial, b—bending.*

The S_e value for high purity aluminum (AA 1050/1100) reported in [11] (34.5 MPa, Table A2) is obtained for purely alternating stress ($R = -1$) conditions and by using an R.R. Moore machine and unnotched rotating-beam specimen. In such a case, a Marin equation [15,38] can be used to adjust the S_e value to the axial loading case by applying load modification factor k_c:

$$S_{e,axial} = k_c S_{e,bending} \tag{A4}$$

where k_c is equal to 0.85 for axial loading [15].

References

[1] Ferro A, Montalenti G, 1964, On the effect of the crystalline structure on the form of fatigue curves, Philos, Mag, 10, 1043. https://doi.org/10.1080/14786436408225410

[2] Ferro A, Mazzetti P, Montalenti G, 1965, On the effect of the crystalline structure on fatigue: Comparison between body-centred metals (Ta, Nb, Mo and W) and face-centred and hexagonal metals, Phil. Mag. J. Theor. Exp. Appl. Phys, 12, 867–875. https://doi.org/10.1080/14786436508218923

[3]   Buck A, 1967, Fatigue properties of pure metals, Int. J. Fract. Mech., 3, 145–152.
       https://doi.org/10.1007/BF00182692

[4]   Grosskreutz J.C, 1971, Fatigue mechanisms in the sub-creep range. ASTM, 495,
       5–60. https://doi.org/10.1520/STP26684S

[5]   Mlikota M, Schmauder S, 2018, On the critical resolved shear stress and its
       importance in the fatigue performance of steels and other metals with different
       crystallographic structures. Metals, 8, 883. https://doi.org/10.3390/met8110883

[6]   Jennings A.T, Burek M.J, Greer J.R, 2010, Microstructure versus Size:
       Mechanical properties of electroplated single crystalline Cu nanopillars. Phys.
       Rev. Lett., 104, 135503. https://doi.org/10.1103/PhysRevLett.104.135503

[7]   Rogne B.R.S, Thaulow C, 2015, Strengthening mechanisms of iron micropillars.
       Philos. Mag, 95, 1814–1828. https://doi.org/10.1080/14786435.2014.984004

[8]   Božić Ž, Schmauder S, Mlikota M, Hummel M, 2014, Multiscale fatigue crack
       growth modelling for welded stiffened panels. Fatigue Fract. Eng. Mater. Struct.,
       37, 1043–1054. https://doi.org/10.1111/ffe.12189

[9]   Monnet G, Pouchon M.A, 2013, Determination of the critical resolved shear stress
       and the friction stress in austenitic stainless steels by compression of pillars
       extracted from single grains. Mater. Lett., 98, 128–130.
       https://doi.org/10.1016/j.matlet.2013.01.118

[10]   Dietmann H, 1991, Einführung in die Elastizitäts- und Festigkeitslehre, Alfred
       Kröner Verlag: Stuttgart, Germany.

[11]   Matweb.com. Matweb—The Online Materials Information Resource. Available
       online:http://www.matweb.com/search/DataSheet.aspx?MatGUID=db0307742df1
       4c8f817bd8d62207368e.

[12]   Islam M.A, Sato N, Tomota Y, 2011, Tensile and plane bending fatigue properties
       of pure iron and iron-phosphorus alloys at room temperature in the air. Trans.
       Indian Inst. Met., 64, 315–320. https://doi.org/10.1007/s12666-011-0064-y

[13]   Fatemi A, Zeng Z, Plaseied A, 2004, Fatigue behavior and life predictions of
       notched specimens made of QT and forged microalloyed steels. Int. J. Fatigue, 26,
       663–672. https://doi.org/10.1016/j.ijfatigue.2003.10.005

[14]   Atzori B, Meneghetti G, Ricotta M, 2011, Analysis of the fatigue strength under
       two load levels of a stainless steel based on energy dissipation. Frattura Integr.
       Strutt., 17, 15–22. https://doi.org/10.3221/IGF-ESIS.17.02

[15]   Budynas R.G, Nisbett J.K, 2015, Shigley's Mechanical Engineering Design, 10th
       ed.; McGraw-Hill Education: New York, NY, USA,.

[16]   Mlikota M, Schmauder S, Božić Ž, 2018, Calculation of the Wöhler (S-N) curve
        using a two-scale model. Int. J. Fatigue, 114, 289–297.
        https://doi.org/10.1016/j.ijfatigue.2018.03.018

[17]   Mlikota M, Schmauder S, 2019, Virtual testing of plasticity effects on fatigue
        crack initiation. In Advances in Engineering Materials, Structures and Systems:
        Innovations, Mechanics and Applications; Zingoni, A., Ed.; CRC Press: London,
        GB, pp. 587–592. https://doi.org/10.1201/9780429426506-102

[18]   Tanaka K, Mura T, 1981, A dislocation model for fatigue crack initiation. J. Appl.
        Mech., 48, 97–103. https://doi.org/10.1115/1.3157599

[19]   Tanaka K, Mura T, 1982, A theory of fatigue crack initiation at inclusions. Metall.
        Trans. A, 13, 117–123. https://doi.org/10.1007/BF02642422

[20]   Mlikota M, Staib S, Schmauder S, Božić Ž, 2017, Numerical determination of
        Paris law constants for carbon steel using a two-scale model. J. Phys. Conf. Ser.,
        843, 012042. https://doi.org/10.1088/1742-6596/843/1/012042

[21]   Glodež S, Jezernik N, Kramberger J, Lassen T, 2010, Numerical modelling of
        fatigue crack initiation of martensitic steel. Adv. Eng. Softw, 41, 823–829.
        https://doi.org/10.1016/j.advengsoft.2010.01.002

[22]   Jezernik N, Kramberger J, Lassen T, Glodež S, 2010, Numerical modelling of
        fatigue crack initiation and growth of martensitic steels. Fatigue Fract. Eng. Mater.
        Struct., 33, 714–723. https://doi.org/10.1111/j.1460-2695.2010.01482.x

[23]   Mlikota M, Schmauder S, Božić Ž, Hummel M, 2017, Modelling of overload
        effects on fatigue crack initiation in case of carbon steel. Fatigue Fract. Eng.
        Mater. Struct., 40, 1182–1190. https://doi.org/10.1111/ffe.12598

[24]   Mlikota M, Schmauder S, 2017, Numerical determination of component Wöhler
        curve. DVM Bericht, 1684, 111–124.

[25]   Božić Ž, Schmauder S, Mlikota M, Hummel M, 2018, Multiscale fatigue crack
        growth modeling for welded stiffened panels. In Handbook of Mechanics of
        Materials, Springer: Singapore, pp. 1–21. https://doi.org/10.1007/978-981-10-
        6855-3_73-1

[26]   Huang X, Brueckner-Foit A, Besel M, Motoyashiki Y, 2007, Simplified three-
        dimensional model for fatigue crack initiation. Eng. Fract. Mech., 74, 2981–2991.
        https://doi.org/10.1016/j.engfracmech.2006.05.027

[27]   Briffod F, Shiraiwa T, Enoki M, 2016, Fatigue crack initiation simulation in pure
        iron polycrystalline aggregate. Mater. Trans., 57, 1741–1746.
        https://doi.org/10.2320/matertrans.M2016216

[28] Božić Ž, Mlikota M, Schmauder S, 2011, Application of the ΔK, ΔJ and $\Delta CTOD$ parameters in fatigue crack growth modelling. Tech. Gaz., 18, 459–466.

[29] Božić Ž, Schmauder S, Mlikota M, 2011, Fatigue growth models for multiple long cracks in plates under cyclic tension based on ΔK_I, ΔJ-integral and $\Delta CTOD$ parameter. Key Eng. Mater, 488–489, 525–528. https://doi.org/10.4028/www.scientific.net/KEM.488-489.525

[30] Jin H, Ko E, Kwon J, Hwang S.S, Shin C, 2016, Evaluation of critical resolved shear strength and deformation mode in proton-irradiated austenitic stainless steel using micro-compression tests. J. Nucl. Mater., 470, 155–163. https://doi.org/10.1016/j.jnucmat.2015.12.029

[31] Krompholz K, Ullrich G, 1985, Investigations into the fatigue crack initiation and propagation behaviour in austenitic stainless steel X5 CrNi 18 9 (1.4301). Materialwiss. Werkstofftech., 16, 270–276. https://doi.org/10.1002/mawe.19850160806

[32] Bao W.P, Xiong Z.P, Ren X.P, Wang F.M, 2013, Effect of strain rate on mechanical properties of pure iron. Adv. Mater. Res., 705, 21–25. https://doi.org/10.4028/www.scientific.net/AMR.705.21

[33] Lorenzino P, Navarro A, Krupp U, 2013, Naked eye observations of microstructurally short fatigue cracks. Int. J. Fatigue, 56, 8–16. https://doi.org/10.1016/j.ijfatigue.2013.07.011

[34] Keil B, Devletian J, 2011, Comparison of the mechanical properties of steel and ductile iron pipe materials. In Pipelines; Jeong, D.H.S., Pecha, D., Eds., American Society of Civil Engineers: Reston, Virginia, USA,; pp. 1301–1312. https://doi.org/10.1061/41187(420)119

[35] Yang L, Fatem, A, 1996, Impact resistance and fracture toughness of vanadium-based microalloyed forging steel in the as-forged and Q&T conditions. J. Eng. Mater. Technol, 118, 71–79. https://doi.org/10.1115/1.2805936

[36] Schijve J, 2009, Fatigue of Structures and Materials, Springer Netherlands: Dordrecht, The Netherlands. https://doi.org/10.1007/978-1-4020-6808-9

[37] Bhaduri A, 2018, Mechanical Properties and Working of Metals and Alloys, Springer Nature: Singapore. https://doi.org/10.1007/978-981-10-7209-3

[38] Marin J, 1962, Mechanical Behavior of Engineering Materials, Prentice Hall: Englewood Cliffs, NJ, USA.

Keyword Index

About the Editor and Authors

PROF. DR. RER. NAT. DR. H. C. SIEGFRIED SCHMAUDER (Author)

Prof. Schmauder was educated in Stuttgart/Germany where he received his Diploma in Mathematics in 1981 at the University and where he finished his PhD in Materials Science in 1988 at the Max-Planck-Institute for Metals Research. Prof. Schmauder then did Postdocs at the University of Tokyo from 1989-1990 and at the University of California at Santa Barbara (UCSB) from 1990-1991.

He returned back to Stuttgart where he became a group leader in Structural Mechanics at the Max-Planck-Institute from 1991-1994. He was nominated as full Professor for Strength of Materials and Materials Techniques in the Faculty of Mechanical Engineering at the University of Stuttgart in 1994.

He has published 6 books and more than 600 scientific papers.

Since 1991 he is the organizer of the annual international workshop on Computational Mechanics of Materials. For more than 10 years he was co-editor of the Journal Computational Materials Science (CMS). His main research field is multiscale materials modelling from atoms to components.

He is a guest professor in Tomsk (Russia) and Tokyo (Japan) and in 2017 he received an honorary doctorate from the Ovidius University in Constanca/Romania for his achievements in multiscale materials modelling.

KIARASH J. DOGAHE, M.SC. (Editor)

Kiarash Dogahe was born in 1993 in Rasht, Iran. He studied Materials Engineering and Nanotechnology at Politecnico di Milano, Italy. Since 2020 he has started to work as a scientific co-worker at the Institute for Materials Testing, Materials Science and Strength of Materials (IMWF), University of Stuttgart, Germany. The main area of his work is the investigation of fatigue behavior of metallic components produced by Additive Manufacturing methods.

DR. MARIJO MLIKOTA (Author)

Dr. Marijo Mlikota is a simulation engineer at the company Magna Steyr Fahrzeugtechnik AG & Co KG in Graz, Austria, where he is a part of the Aerospace department. He was born in Konjic, Bosnia and Herzegovina, in 1986. He obtained his M.Sc. degree in mechanical engineering, with specification in computer aided engineering, in 2010 at the Faculty of Mechanical Engineering and Naval Architecture,

University of Zagreb, Croatia. From 2011 until 2021 he worked as a scientific co-worker at the Institute for Materials Testing, Materials Science and Strength of Materials (IMWF), University of Stuttgart, Germany. He obtained his doctoral degree there in 2020, too. Aside of fatigue of metallic materials, he was involved in the research field of metal erosion and soft tissue behavior.

PROF. DR. SC. ŽELJKO BOŽIĆ (Author)

Prof. Željko Božić is a Full Professor at the Faculty of Mechanical Engineering and Naval Architecture, University of Zagreb, where he is the Head of the Department of Aeronautical Engineering. He achieved a Dipl. Ing. degree in mechanical engineering in 1987 and a M.Sc. degree in structural mechanics in 1992 at the Faculty of Mechanical Engineering and Naval Architecture, University of Zagreb.

In 1993 he was awarded the Japanese Government (Monbusho) Scholarship for his PhD study in the Department of Naval Architecture and Ocean Engineering at Yokohama National University in Japan, where he received his PhD in the Fatigue and Fracture of Ship Structures, in 1997. He was awarded the Silver Medal of Yokohama National University. From 2001 to 2004 he worked in R&D Department in ALSTOM Power, Baden, Switzerland, on fatigue and integrity of gas turbine structures. He was a DFG (German Science Foundation) guest professor at the University of Stuttgart in 2010, 2011, 2012 and 2014. He was a visiting professor at University of Bergamo, Italy, in 2018 and 2021.

His research interests include analytical, numerical and experimental methods to treat fatigue and fracture of structures. In particular, modelling and analysis of fatigue crack propagation considering residual stress effects and interaction of multiple cracks. He is the European Structural Integrity Society (ESIS) Executive Committee (ExCo) member and he was the Organizing Committee Chair of the 4th ESIS Summer School, held in Dubrovnik in 2017. He was the Organizer and Chair of several International Conferences including the International Conference on Structural Integrity and Durability (ICSID), which he initiated in 2017 and which is since then held every year in Dubrovnik. He was Guest Editor of Special Issues of the Engineering Failure Analysis journal, International Journal of Fatigue, Fatigue and Fracture of Engineering Materials and Structures, Procedia Structural Integrity, and other journals. He gave more than 70 oral presentations at conferences and workshops including five plenary lectures.

www.ingramcontent.com/pod-product-compliance
Lightning Source LLC
Chambersburg PA
CBHW071502210326
41597CB00018B/2661